Lecture Notes in Earth Sciences

Edited by Gerald M. Friedman, Horst J. Neugebauer
and Adolf Seilacher

3

Thomas Aigner

Storm Depositional Systems
Dynamic Stratigraphy in Modern and Ancient
Shallow-Marine Sequences

Springer-Verlag
Berlin Heidelberg New York Tokyo

Author

Dr. Thomas Aigner
Universität Tübingen
Institut und Museum für Geologie und Paläontologie
Sigwartstraße 10, D-7400 Tübingen, FRG

ISBN 3-540-15231-8 Springer-Verlag Berlin Heidelberg New York Tokyo
ISBN 0-387-15231-8 Springer-Verlag New York Heidelberg Berlin Tokyo

© by Springer-Verlag Berlin Heidelberg 1985
Printed in Germany

Printing and binding: Beltz Offsetdruck, Hemsbach/Bergstr.
2132/3140-543210

"Nature vibrates with rhythms, climatic and
diastrophic, those finding stratigraphic expression
ranging in period from the rapid oscillation of
surface waters, recorded in ripple-marks, to those
long-deferred stirrings of the deep imprisoned titans
which have divided earth history into periods and
eras. The flight of time is measured by the weaving
of composite rhythms - day and night, calm and storm,
summer and winter, birth and death - such as these
are sensed· in the brief life of man ...
... the stratigraphic series constitutes a record,
written on tablets of stone, of the lesser and
greater waves of change which have pulsed through
geologic time".

JOSEPH BARRELL (1917)

PREFACE

It was only during the last few years, that the geological effects of
storms and hurricanes in shallow-marine environments have been better
appreciated. Not only were storm deposits recognized to dominate many
shelf sequences, they also proved to be valuable tools in facies and
paleogeographical analysis. Additionally, storm layers form important
hydrocarbon reservoirs.

Storm-generated sequences are now reasonably well documented in terms
of their facies associations in the stratigraphic record. Much less is
known, however, about the effects and the depositional processes of
modern storms, and about the styles of storm sedimentation on basin-
wide scales. Accordingly, the goal of this study is two-fold:

1. it presents two case studies of modern carbonate and terrigenous
clastics storm sedimentation. The models derived from these actualistic examples can be used to interprete possible ancient analogues.

2. it presents a comprehensive analysis of an ancient storm depositional system (Muschelkalk) on a basin-wide scale.

The underlying approach of this study is a process-oriented analysis
of sedimentary sequences, an approach that was summarized by Matthews
(1974, 1984) as "dynamic stratigraphy". The integration of actualistic models with a "dynamic" stratigraphic analysis helps to understand the dynamics of storm depositional systems; these models have a
potential to be applied to other basins and to predict the facies
organisation and the facies evolution in such systems.

ACKNOWLEDGEMENTS

First and foremost I would like to sincerely thank Prof. Dr. A. Seilacher. Over the years of my studies in Tübingen he acted as an adviser in an always pleasant and fruitful atmosphere, was a constant source of guidance and encouragement, and helped me sharpening my own ideas.

To come up with a doctoral dissertation is never alone a product of oneself. There are always many people and incidents along the way that have helped oneway or the other. I should start thanking my parents who have always generously supported my passion for rocks and fossils. Then there were school teachers, notably H. Fischer and H. Huber, who directed my attention from collecting stones to the geosciences. H. Hagdorn has always been a stimulating and invaluable Muschelkalk compagnion.

Contacts with geological friends, fellow students, faculty and scholars from abroad have had an immense impact on me, as have opportunities to travel. One year of studies in sedimentology with Dr. R. Goldring and Prof. J.R.L. Allen at the University of Reading/England has been a key experience. Since I left Reading, Roland continued to try teaching me how to write in English in reviewing many papers and manuscripts. The opportunity to do a Diplom-Thesis in Egypt, also supervised by Prof. Seilacher, and successively to join the Sphinx Project of the American Research Center in Egypt were scientific and personal experiences I do not want to miss. During several stays at the Senckenberg-Institute in Wilhelmshaven, Prof. Dr. H.-E. Reineck has most generously taught me principles of marine geology and supervised the North Sea work included in this dissertation. Two expeditions with Prof. Dr. J. Wendt into the Moroccan Sahara were sometimes hot and dry, but they were always most educating - and fun.

I had the fine opportunity to study modern carbonate environments of South Florida during a 9-month's stay at the University of Miami, where Dr. H.R. Wanless acted as a most generous and stimulating supervisor who had always time for me. Many friends in Miami helped me battling against the hazards of marine work such as weather, boat problems, coring etc. Notably I want to thank V. Rossinski, J. Meeder, P. Harlem, M. Almasi, R. Parkinson, F. Beddour, and A. Droxler. Dr. R.N. Ginsburg kindly allowed me to use facilities at Fisher Island Station. The Rosenstiel School and the Senckenberg-Institute also provided technical assistance.

Among the many colleagues that discussed problems with me or guided me through their field areas, I want to particularly thank Dr. R. Bambach, Dr. J. Bourgeois, Dr. P. Duringer, Dr. F. Fürsich, Dr. R. Goldring, H. Hagdorn, A. Hary, Dr. S. Kidwell, Dr. R. Mundlos, and Dr. A. Wetzel. Dr. R. Hatfield made some radiocarbon determinations, Prof. Dr. H. Friedrichsen some isotope analysis, Prof. Dr. C. Hemleben provided access to a word processor, H. Hüttemann helped with the SEM. Particularly I thank W. Ries who helped much with rock cutting and thin sections and W. Wetzel who made much of the photographs and reproductions. Financial support from the Deutsche Forschungsgemeinschaft (SFB 53) is also gratefully acknowledged.

Prof. Dr. A. Seilacher, Dr. R. Goldring, Prof. Dr. G. Einsele, Prof. Dr. H.-E. Reineck and Prof. Dr. J. Wendt looked through the original version of the manuscript, H. Hagdorn checked parts on crinoidal limestone. Dr. W. Engel and the Springer-Verlag is thanked for publishing my thesis as it stands in the LECTURE NOTES series. I am most grateful to all those who made this work possible.

S U M M A R Y

This study comprises (1) two case histories of storm sedimentation in modern shallow-marine environments, and (2) a comprehensive analysis of an ancient storm depositional system on a basin-wide scale. These examples are understood as a contribution to "dynamic stratigraphy", the process-oriented analysis of sedimentary sequences, and are believed to be applicable to other shallow-marine basins.

Basic physical processes during storm sedimentation involve, according to a general model of Allen (1982, 1984), three main categories:

a) barometric effects due to gradients in atmospheric pressure leading to raised water levels at the shore (coastal water set-up).

b) wind effects cause (1) onshore wind drift currents in nearshore surface water, which are compensated by (2) offshore oriented bottom return flows (gradient currents).

c) wave effects set up oscillatory bottom flows; superimposed unidirectional flows lead to combined storm flows.

Sedimentary responses to storm processes in modern shallow-marine environments of South Florida and the North Sea support this general model. Storm effects in shallow, nearshore water are dominated by onshore directed wind drift currents. These cause the formation of onshore sediment lobes in nearshore skeletal banks of South Florida. Successive hurricane-generated "spillover lobes" contribute as depositional increments to episodic and relatively rapid accretion and buildup of non-reef skeletal banks that coarsen and are increasingly winnowed upwards. Similar episodic buildup can be inferred for near-shore bioclastic banks in the fossil record.

Responses to storm processes in offshore shelf areas such as the German Bay (North Sea) involve seaward transport of sands and shells from coastal sand sources by offshore flowing bottom currents (gradient currents) and their deposition as offshore storm sheets (tempestites). Qualitatively and quantitatively, such tempestites show systematic changes in their sedimentological and paleoecological characteristics from nearshore to offshore. These proximality trends reflect the decreasing effect of storms away from the coastal sand source and with increasing water depth.

A large variety of storm responses, involving patterns found in actualistic analogues, are revealed by a basin-wide analysis of the Upper Muschelkalk (M. Triassic, SW-Germany), an intracratonic ancient storm depositional system. In this setting, dynamic processes are reconstructed based on a hierarchical three-level stratigraphic analysis:

1. At the lowest level, individual strata record episodic storm events operating on a gently inclined carbonate ramp system. Paleocurrents suggest alongshore winds and storm tracks from the Tethys to the NE into the German Basin. Similar to actualistic models (Swift et al., 1983), these are likely to have induced combined oscillatory/unidirectional geostrophic bottom currents in offshore areas (distal tempestites). At the same time, longshore wind stress will drive surface water landward (Coriolis effect), causing landward sediment transport and accumulation of nearshore skeletal banks, in a fashion similar to modern examples from South Florida. Coastal water set-up is compensated by offshore directed bottom return flows, much like

gradient currents in the present-day North Sea. These backflows erode surge channels, through which sediment is funneled offshore to become deposited as proximal tempestites.

2. At an intermediate level, storm beds in the Upper Muschelkalk tend to be arranged cyclically into 1-7 m thick coarsening- and thickening-upwards facies sequences, that record an upward transition from distal to proximal tempestites, i.e. progressive shallowing. Different types of asymmetrical coarsening-upward cycles also show systematic changes in the molluscan and trace fossil associations that reflect a change in substrate conditions. Widespread changes from soft into firm and shelly substrates allowed in several instances for virtually instantaneous and geographically widespread colonisation of cycle tops by specific brachiopods and crinoids. The massive, often amalgamated, condensed and "ecologically fingerprinted" tops of such cycles (e.g. Spiriferina-Bank, Holocrinus-Bank, see Hagdorn, 1985) serve as principal marker beds. Similarly, prominent marlstone horizons have long been used in lithostratigraphic correlation ("Tonhorizont alpha, beta etc."). Genetically, the marlstone horizons represent the transgressive bases, while massive units are the regressive tops of minor transgressive/regressive cycles.

3. At a still higher level, vertically stacked coarsening-upward cycles constitute a still larger overall cycle forming the entire Upper Muschelkalk. This overall transgressive/regressive cycle is comparable in thickness and duration to the "third-order cycles" of Vail et al. (1977) and corresponds to a large-scale late Anisian/Ladinian transgressive/regressive cycle (Brandner, 1984), which is likely to be eustatically controlled. On the other hand, the distribution of minor cycles and the general organisation of the South-German Basin corresponds well to the underling Variscan structural zones. The "marginal" facies zones correspond to the Moldanubikum in the SE and the Rhenoherzynikum in the NW, while the more rapidly subsiding "central" facies zone is situated ontop of the Saxothuringikum. Within the Moldanubian structural zone, minor cycles can be easily correlated over several ten's of km, but cycle patterns change in character in the adjacent structural zones and are often difficult to correlate. It thus appears that the South-German intracratonic basin expresses the sutures of a former continental collision and that basin dynamics is controlled by an interplay of eustatic as well as structural movements.

In conclusion, an integration of actualistic models with a "dynamic" stratigraphic analysis allows a better understanding of storm processes and their depositional products and provides a base to predict facies patterns over a range of shallow-water environments. Moreover, "dynamic stratigraphy" as outlined here is a tool to reconstruct processes in shallow-marine basins, moving from the smallest (individual strata) to larger levels (whole basin sequence).

C O N T E N T S

Part I

MODERN STORM DEPOSITIONAL SYSTEMS:

ACTUALISTIC MODELS

1. GENERAL PROCESSES

OF STORM SEDIMENTATION

In shallow-marine rock sequences, "wave base" has long been inferred to control sedimentation (e.g. Barrell, 1917; Irwin, 1965). In addition, studies in modern shallow-marine environments have shown that during storms, waves are able to affect the shelf bottom well below "normal" or "fair-weather" wave base (e.g. Komar et al., 1972).

Thus the "storm versus fair-weather concept" has been widely applied to modern and ancient shelf deposits (for a review see Johnson, 1978). Accordingly, facies models for shallow-marine sequences commonly distinguish "fair-weather wave base" and "storm wave base" (e.g. Wilson, 1975; Hamblin & Walker, 1979; Walker, 1980b; Einsele & Seilacher, 1982). Hamblin & Walker (1979) first proposed a distinct zonation of storm-generated facies types with "hummocky cross-stratification being formed between "fair-weather" and "storm wave base" by density currents. Their concepts have initiated considerable discussion in the literature (e.g. Morton, 1981; Nelson, 1982; Dott & Bourgeois, 1982; Swift et al., 1983).

However, on modern shelves, the depth of wave-reworking shows strong seasonal variations: wave base during winter storms may be twice as deep as during the summer (Komar et al., 1972; Aigner & Reineck, 1983) Thus there seems to be a continuous spectrum of conditions, and the terms "fair-weather" and "storm wave base" are probably rather artificial categories, which will therefore not be used here.

The basic physical processes during storm sedimentation have been illustrated by Allen (1982, 1984) in a comprehensive model. Since elements of this model are readily applicable to our two case studies of modern storm sedimentation, the fundamentals of Allen's model will first be briefly reviewed.

Allen's (1982, 1984) model assumes simple laminar flows of a storm approaching the coastline at a perpendicular angle and with constant speed and direction. For the sake of simplicity, complications due to the effects of tides and the Coriolis-force are not considered. Three basic categories of processes and effects can be distinguished during storm sedimentation (Fig. 1):

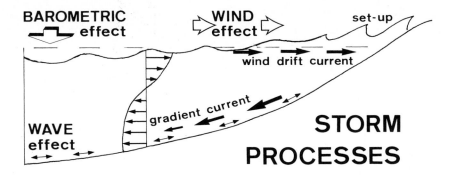

Fig. 1. The complexity of phenomena during storm events can be simplified by distinguishing three main categories of physical processes: (1) <u>Barometric effects</u> causing coastal water set-up; (2) <u>wind effects</u> resulting in onshore directed wind drift currents in surface waters, that are compensated by offshore directed gradient currents in bottom waters; (3) <u>wave effects</u> that mobilize bottom sediment and makes it available for lateral transport. Simple case of onshore blowing storm, interactions with tides and the Coriolis effect not considered.

1. <u>Barometric effect</u>. Cyclonic depressions are accompanied by a horizontal gradient of atmospheric pressure, thus raising the water level at the shore (coastal set-up). Since a pressure difference of one millibar corresponds to a difference in surface elevation of about one cm, typical cyclones raise the water level at the coast for about 1/2 m.

2. <u>Wind effects</u> involve the combination of two processes:

a) The drag of the onshore blowing wind not only further contributes to coastal water set-up, but also results in a nearshore <u>wind-drift current</u> that acts in the same (onshore) direction.

b) The tilt in the water surface due to barometric effect and wind-drift current is compensated by a near-bottom return flow, the <u>gradient current</u>, that flows opposite to wind drift (offshore). Thus the combined motion in a water body is an onshore near-surface flow and a compensating offshore bottom return flow. Consequently, predominanty onshore sediment transport would be expected in shallow, nearshore water (due to the wind-drift current), in contrast to deeper waters, where offshore sediment transport should prevail.

3. _Wave effects_ cause near-bed oscillatory flows that are responsible for stirring-up and mobilizing bottom sediments.

A combination of wave-stirring with unidirectional wind-induced gradient currents would cause pulsating _combined_ _flows_, that transport nearshore sediments offshore and deposit them as graded storm layers.

The following two chapters on modern storm sedimentation could be viewed as a test of Allen's (1982, 1984) model. Both examples indeed support his concept:

A) Storm sedimentation in _nearshore_ carbonate banks of South Florida mainly involves _onshore_ sediment transport (due to onshore wind drift currents).

B) Storm sedimentation in the _offshore_ terrigenous clastics shelf of the German Bay is dominated by _offshore_ sediment transport (due to the offshore flowing gradient current).

These two basic mechanisms can also be recognized in an ancient storm depositional system (M. Triassic Muschelkalk) that is analyzed in the second part of this study. In that example, however, the simple mechanisms outlined here are modified and somewhat complicated by (1) alongshore storms, and (2) interaction with the Coriolis effect.

2. STORM SEDIMENTATION
IN NEARSHORE SKELETAL BANKS,
SOUTH FLORIDA

2.1. Introduction

The effects of storms in nearshore environments involve (1) coastal erosion by wave attack, (2) raised water levels (wind set-up), and consequently (3) dominantly landward sediment transport caused by onshore wind-drift currents. Landward sediment transport is familiar in supratidal storm layers (e.g. Shinn, 1983), and in washover fans commonly associated with barrier island systems (e.g. Nummedale et al., 1980). Similarly, onshore oriented spillover lobes were found to be important for the development of oolite sand "barriers" and banks in the Bahamas (Ball, 1967; Hine, 1977). Much less is known, however, about the effects of storms in other carbonate "sand bodies", such as skeletal banks, although such skeletal buildups are very common in modern nearshore carbonate environments and in many ancient shallow-marine carbonate sequences.

The object of this chapter is to

1. provide an actualistic example of storm sedimentation in nearshore skeletal banks and to document storm-generated sequences not known before from such settings;

2. to examine the role of storms in the development and growth of near-shore skeletal banks as a style of storm sedimentation that contrasts with more offshore level-bottom environments.

This chapter is based on work at the Rosenstiel School of Marine and Atmospheric Sciences of the University of Miami where it was super-vised by Dr. H. Wanless.

2.2. Methods

Aerial photographs of the Safety Valve skeletal bank complex (Biscayne Bay) covering the last 40 years were examined, particularly to detect changes associated with hurricanes. Effects caused by a strong storm on January 20, 1983 were directly studied from a small plane and on the ground. The surface sediments and their biota were briefly surveyed. In the Safety Valve, cores from 30 localities were taken, partly as handpush cores on the bank surface (for method see Wanless, 1969), partly as vibrocores in slightly deeper water. In the Sandy Key area (Florida Bay), 15 handpush cores were taken. A selection of cores was impregnated with polyester resin (for method see Ginsburg et al., 1966) and partly X-radiographed. Thin sections were also made from representative sections. In order to study the faunal content, skeletal layers were sieved through a 2 mm sieve, and the remaining skeletal materials studied for genera present and for preservation. Radiocarbon age determinations have been made from 3 skeletal samples.

2.3. Study area and previous work

The Safety Valve belt of skeletal banks is situated southeast of Miami, Florida, between Key Biscayne and Soldier Key. It forms a shallowly submerged barrier between Biscayne Bay and the inner Florida shelf (Fig. 2A,B). The axis of this belt strikes north-south, parallel to a now submerged ridge of Pleistocene Key Largo Limestone, which contours the eastern margin of the belt (Fig. 2A,B). Ball (1967) refers to the Safety Valve as a small "tidal bar belt". The tidal bar belt of the Safety Valve is about 3-4 km broad and 8-9 km in length. Numerous tidal channels dissect it into elongated banks ("bars") extending perpendicularly to the axis of the belt (Fig. 2B). Individual banks vary in width from a few decameters to several hundred meters, while banks in the northern part are generally wider than the ones in the south. The bank surfaces are variable through time. Seagrass (Thalassia) commonly forms dense covers, but bare substrata made of coarse skeletal material may also be present, particularly after severe storms and hurricanes (Warzeski, 1967).

Wanless (1969) has given a detailed description of the geomorphological features in the area as well as a reconstruction of the depositional history in response to the late Holocene rise of sea level.

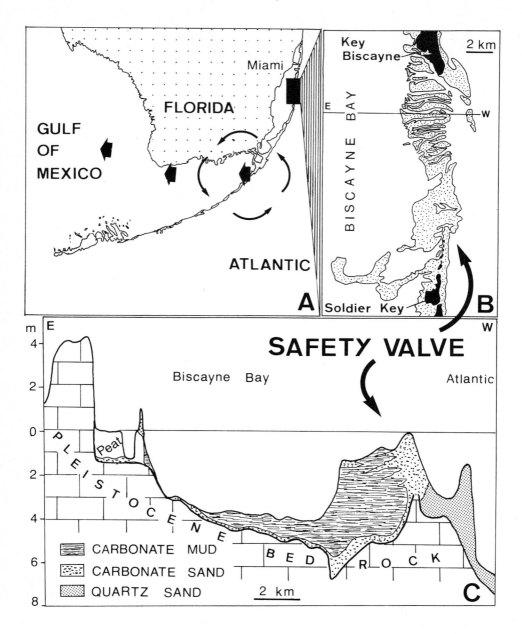

Fig. 2. General setting of "Safety Valve" study area. A) Location of the Safety Valve tidal bar belt east of Miami (black), between the Atlantic and Biscayne Bay. Arrows indicate path and wind circulation across the center of Huricane Betsy, after Perkins & Enos (1968). B) Overview of Safety Valve banks (stippled: shallowly submerged banks). C) East-West transect from the mainland (Miami) - Biscayne Bay - Safety Valve, simplified after Wanless (1969). Note ridge of bedrock (Key Largo Limestone) localizing the Safety Valve banks.

Ginsburg & James (1974) included the Safety Valve in their review on "algal banks" of South Florida and distinguished two communities : (1) a <u>Porites/Goniolithon</u> community on windward bank margins, and (2) a banktop community with <u>Thalassia</u>, <u>Halimeda</u>, molluscs etc.

For comparison, the "Sandy Key" area of skeletal banks separating the Gulf of Mexico from Florida Bay was also surveyed (Fig. 3). This area was affected by Hurricane Donna and has been briefly described by Ball (1967). It represents a good modern example of a carbonate ramp system.

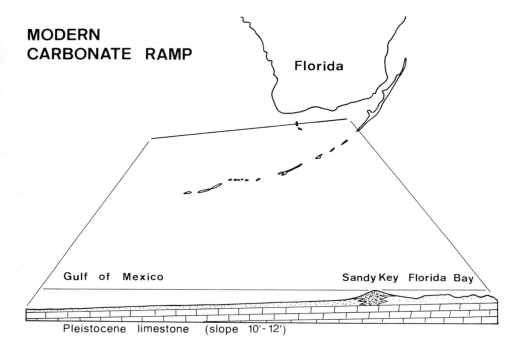

MODERN CARBONATE RAMP

Florida

Gulf of Mexico   Sandy Key Florida Bay

Pleistocene limestone (slope 10'-12')

<u>Fig. 3.</u> General setting of "Sandy Key" study area, a belt of skeletal banks and shell islands on a gently inclined ramp separating the restricted area of Florida Bay from the open-marine areas towards the Gulf of Mexico, where only a very thin veneer of modern sediment overlies Pleistocene bedrock (highly schematic). Water depths range between 1/2 m to a few meters in Florida Bay to some tens of meters and more towards the Gulf.

2.4. Geomorphology of Safety Valve banks

Four major geomorphological elements can be distinguished in the
Safety Valve banks (Fig. 4):

BANK GEOMORPHOLOGY

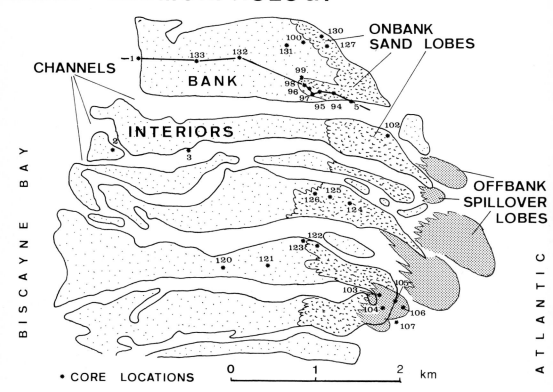

Fig. 4. Major geomorphological elements of the Safety Valve: (1) tidal
channels, (2) offbank spillover lobes seaward of tidal channels, (3)
onbank sand lobes on seaward bank margins, and (4) bank interior
areas. Numbers represent core locations, east-west transect is shown
in Fig. 6, that also gives an orientation on water depths.

1. Channels cutting the main body of the banks. They are generally 2-4
m deep, fairly straight, but tend to bifurcate on the bayward end.
Wanless (1969) noted that many of the channels were formed by storms
breaching the banks.

2. _Offbank spillover lobes_. Prominent lobes of skeletal material off the seaward end of several tidal channels are 2oo-1ooo m in length and 2oo-5oo m in width. Wanless (1969) concluded that these spillover lobes formed during storms.

3. _Onbank sand lobes_. From the seaward portions of the bank flats, a zone of coalescent lobes of skeletal sand and gravel extends up to about 1000 m onto the banks (Fig. 4), each lobe being 5o-200 m in width. As documented below, these lobes are also formed during storms and hurricanes.

4. The _bank interiors_ of the Safety Valve contrast by their muddiness with the coarse seaward margin and include elongated deeper "ponds", that may represent remnants of infilled abandoned channels.

2.5. Sedimentary facies and bank stratigraphy

2.5.1. Coralgal pack- to grainstone (Fig. 5A)

This facies occurs along some channel margins, but mainly forms the prominent onbank sand lobes. It is characterized by medium sand to gravel-sized fragments of _Porites_, _Halimeda_, molluscs and some _Goniolithon_ in a mostly clean-washed grain-supported texture. Gravel-sized bioclasts such as coral fragments or large mollusks may be tightly packed together, and mark the basal lag of fining-up sequences that often overlie erosion surfaces (as evidenced by truncated sea-grass rootlets). In the bank stratigraphy, these lobes form wedge-shaped depositional units (Fig. 6). Their thickness decreases from the bank margins into the bank, while their mud content increases.

2.5.2. _Halimeda_ packstone (Fig. 5B)

This facies is primarily composed of fine sand to fine gravel sized particles of _Halimeda_ with some mollusc shells and contains only minor amounts of corals or _Goniolithon_. Imbrication or orientation of _Halimeda_ plates parallel to bedding is common. _Halimeda_ packstone occurs as thin layers within the bank interior, and along bank margins adjacent to channels, but is most characteristic of offbank spillover lobes. Vibrocores from spillover lobes on the seaward side of a tidal

channel (see Fig. 4) show a layered stratigraphy (Fig. 7). Close to
the channel mouths, vibrocores display amalgamated well-sorted
Halimeda pack- to grainstone, where layering is clearly marked by
truncated seagrass roots (Fig. 9E), skeletal lags and overall
fining-up sequences. In an offshore direction, the Halimeda layers
interfinger with wackestone or quartz-carbonate sand layers. At the
same time they decrease in thickness and sorting, while the mud
content increases.

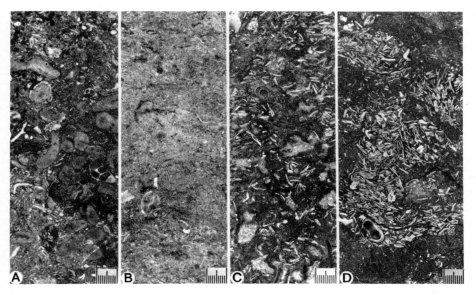

Fig. 5. Main types of sedimentary facies (impregnated slabs from sediment cores). A) Coralgal grainstone with abundant Porites fragments
from storm-generated sediment lobe on seaward bank margin. B) Well-
sorted Halimeda pack- to grainstone from offbank spillover lobe. C)
Pellet-rich Halimeda-mollusc wackestone from the bank interior facies.
D) Burrow filled with Halimeda packstone.

2.5.3. Pellet-rich Halimeda-mollusc wackestone (Fig. 5C,D)

This facies predominates in the interior parts of the banks. Because
of very intense bioturbation, Halimeda plates are mostly chaotically
oriented and/or occur in burrow fills (Fig. 5D). Seagrass rootlets
penetrating the sediment are most common in this facies, and evidence
for physical layering is very difficult to recognize.

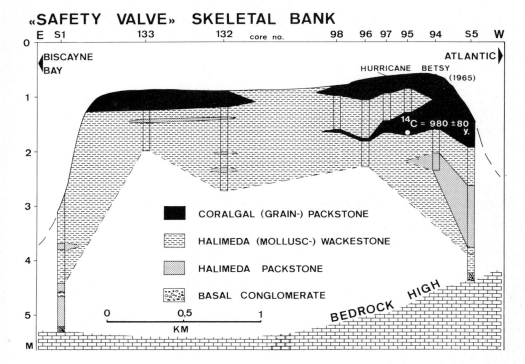

Fig. 6. Distribution of sedimentary facies in core transect across northermost bank in the Safety Valve (see Fig. 4). Note wedges of coralgal pack- to grainstone (black) extending from seaward bank margin into the bank interior.

2.5.4. Mollusc wacke- to packstone with lithoclasts

The first few centimeters or decimeters of sediment overlying Pleisto- cene bedrock consist of molluscan wacke- to packstone with appreciable amounts of quartz sand, reworked and partly blackened limestone bed- rock lithoclasts, as well as blackened shells. Halimeda plates are rare. This facies seems to record the first phase of marine sediment- ation following the Holocene transgression over Pleistocene bedrock (transgression lag).

2.5.5. Quartz sand

Beds of quartz sand, a few cm to several dm in thickness and with variable amounts of intermixed carbonate skeletal sand occur both in onbank sand lobes and in offbank spillover lobes. Well-developed

OFFBANK SPILLOVERS

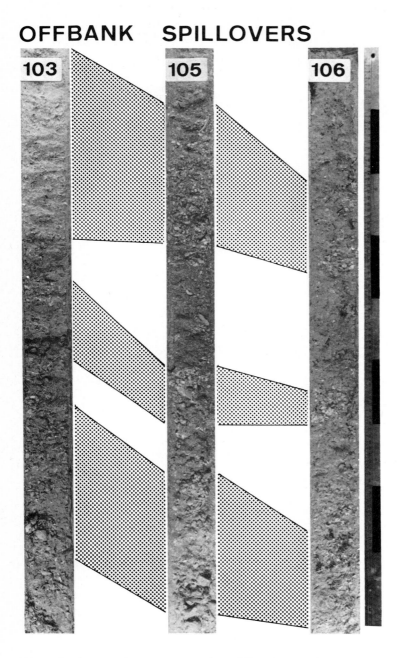

Fig. 7. Layered sequences in offbank spillover lobe (see Fig. 4 for location). Note that skeletal layers interbedded with muddier intervals become thinner, muddier and less distinct away from the channel mouth (towards the right).

fining-up sequences (Fig. 9B), together with their exotic occurrence within a carbonate bank complex indicate that this sand was laterally introduced, most likely during high-energy events (storms). The most obvious source area is a submarine shoal extending from the barrier island of Key Biscayne, about 1-2 km east (seaward) of the Safety Valve.

2.6. Evidence for storm sedimentation

Geomorphologic, stratigraphic and biostratinomic data provide evidence for the effects of storms and hurricanes in the Safety Valve skeletal banks.

2.6.1. Geomorphologic evidence

Day-to-day current velocities in the Safety Valve are not capable of transporting gravel-sized skeletal material (Ball, 1967). The sand lobes and spillover lobes must therefore result from episodic pulses of a higher flow regime during which coarse material could be entrained.

Offbank spillover lobes are similar to those in Bahamian mobile oolite banks, for which Ball (1967) and Hine (1977) concluded that they form during intermittent high-energy events such as severe storms and hurricanes and are only surficially reworked and modified by daily tidal processes. Possible mechanisms in the Safety Valve include (1) storm return flows flushing in an offshore direction through the Safety Valve channels and compensating for water set-up in Biscayne Bay during onshore (north-moving) storms and hurricanes, and (2) southerly drift currents caused by south-moving winter storms.

Onbank sand lobes are similar to washovers associated with clastics barrier islands. Such washovers are generally attributed to exceptional storms (e.g. Nummedale et al., 1980). An analogous origin can be inferred for the Safety Valve sand lobes. Direct proof for a hurricane origin comes from a series of aerial photographs covering the time period between 1940 and 1972 (Fig. 8). Barren layers of skeletal sand and gravel forming lobes only appear after major storms and hurricanes, especially after Hurricane Betsy (1965) which passed

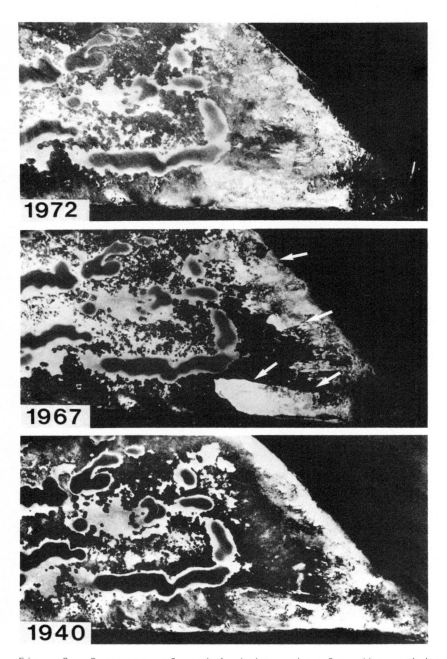

Fig. 8. Sequence of aerial photographs of northermost bank in Safety Valve. Bottom: photograph taken in 1940, prior to Hurricane Betsy. Middle: photograph taken in 1967, after Hurricane Betsy; note new skeletal sand lobes extending from the seaward bank margin far into the bank interior (arrows). Top: photograph from 1972 showing beginning recolonisation of the Betsy sand lobe by seagrass and some new sand lobes. Width of view approximately 2 km.

directly through this area (see Fig. 2A). These skeletal lobes may either cover or erode previously vegetated bottoms. The 60 miles/hour winter storm of January 2o, 1983 has reactivated the Betsy sand lobe in remobilizing the surficial layers of skeletal sand and smothering parts of its seagrass cover. This suggests that a number of storms can be amalgamated in one sand lobe.

Other skeletal layers, similar to the Betsy-layer in composition and in wedge-shaped geometry, have been recorded at deeper levels within the bank stratigraphy(Fig. 6). In analogy with the surface layer, these are interpreted as sand lobes related to hurricanes earlier in the bank history.

2.6.2. Stratigraphic evidence

1. Sharp-based skeletal layers. (Fig. 9, 10). Densely packed concentrations of skeletal material (mostly mollusc-Halimeda-Porites packstone, some grainstone) ranging in thickness from a few cm to several dm were commonly recorded in the sand lobes of the seaward bank portions and in deeper parts of sediment cores. These layers trun- cate underlying seagrass rootlets and/or rhizomes, indicating erosion (Fig. 9). Erosion is most likely due to storms and hurricanes; Ball et al. (1967) reported extensive erosion and exposed seagrass rootlets on Florida Bay mudbanks after Hurricane Donna. Sharp-based skeletal layers can therefore be interpreted as the result of storm erosion followed by physical accumulation of bioclastic particles.

2. Fining-up sequences. Several types of fining-up sequences are recognized in surface sand lobes as well as in sediment cores and are illustrated in Figs. 9 and 11. Basal erosion in all types is indicated by truncation of seagrass roots or by the position of the erosion sur- face directly on a rhizome horizon that normally lies 1o-2o cm below the sediment surface. Bioturbation and seagrass become increasingly more abundant upwards in the fining-up sequences, suggesting re- colonization of the graded unit. Ball et al. (1967) also described "graded layers of skeletal sand" deposited onto banks during Hurricane Donna. Onshore (bayward) sediment transport can be documented in that Goniolithon and Porites fragments are swept from the bank margins into the bank interiors. Also, a wedge-shaped geometry of onbank sand lobes (and of offbank spillover lobes) implies onshore (and some offshore) sediment transport.

SEQUENCES OF PHYSICAL EVENTS

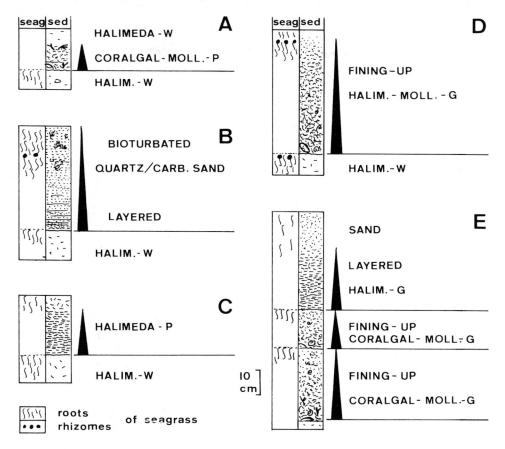

Fig. 9. Some common types of sedimentary sequences found in the Safety Valve banks that indicate physical sedimentation. A) Sharp-based (truncated seagrass roots) unit of coralgal-mollusc packstone within Halimeda wackestone. B) Sharp-based (truncated seagrass roots) unit of layered quartz/carbonate sand with renewed bioturbation and seagrass colonisation at the top. C) Sharp-based fining-upward sequence of Halimeda packstone with Halimeda plates being oriented parallel to bedding. D) Sparp-based fining-upward sequence of Halimeda-mollusc grainstone. Post-event recolonisation is indicated by seagrass rhizomes at the top of the sequence. E) Amalgamation of skeletal layers as indicated by two horizons with truncated seagrass roots. W = wackestone, P = packstone, G = grainstone.

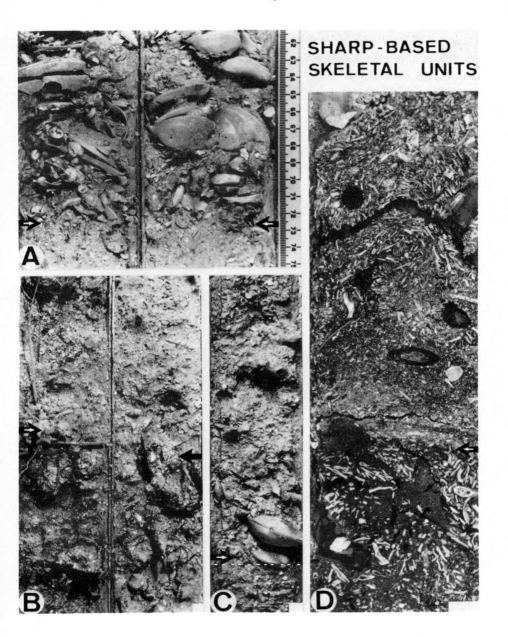

Fig. 10. Examples of sharp-based (arrows) skeletal units that indicate physical events in carbonate bank buildup. A) Thin layer of densely packed molluscan shell material, including double valved Codackia (core no. 131). B) Core (no. 130) from onbank sand lobe showing skeletal sand overlying truncated seagrass roots and rhizomes. C) Core (no. 94) from onbank sand lobe showing layer of skeletal gravel with basal imbrication of double-valved Codackia shells (the larger one was used for radiocarbon age determination). D) Sharp-based unit of Halimeda pack- to grainstone, erosively overlying Halimeda wackestone (specimen courtesy of H.R. Wanless). Scales in B,C,D = 1 cm.

FINING-UP SKELETAL UNITS

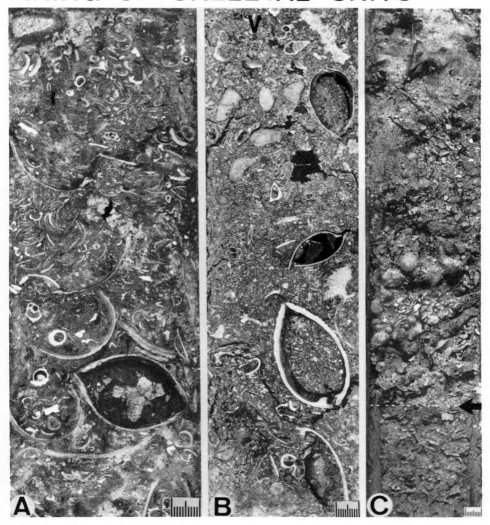

Fig. 11. Fining-upward skeletal units indicating carbonate bank accretion by physical events. A) Fining-up molluscan packstone; note preferred convex-down orientation of large pelecypod shells, indicating rapid dumping (core no. S3). B) Fining-up mollusc-coralgal packstone; note abundance of double-valved pelecypods (core no. 100). C) Sharp-based unit of fining-up coralgal grainstone from onbank sand lobe; note basal lag of Porites fragments overlain by skeletal sand (core no. 127). Scales = 1 cm.

2.6.3. Biostratinomic evidence

(1) <u>Preservation of fauna</u>. A striking aspect in skeletal layers is the abundance of well-preserved, articulated, double-valved pelecypods such as <u>Chione cancellata</u> and <u>Codackia orbicularis</u> (Fig. 10A,C; 11A,B). This strongly suggests rapid winnowing and immediate redeposition during one instantaneous event rather than slow accumulation. The fining-up sequences are therefore the product of episodic erosion (truncated seagrass roots), followed by rapid redeposition of sediment (double-valved pelecypods) during the waning stages of the event.

PRESERVATION OF FAUNA IN SPILLOVER LAYER

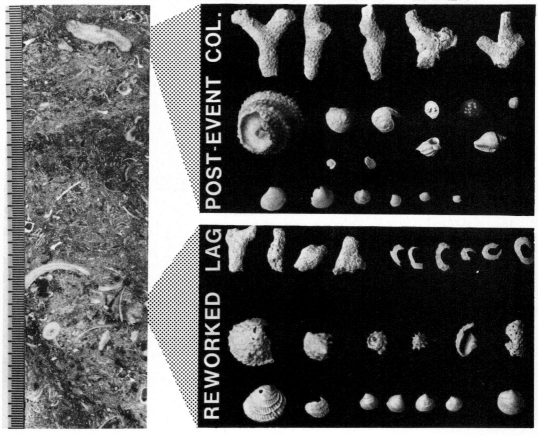

Fig. 12. Differences in composition and preservation at the base and at the top of a spillover layer. Note especially the extremely worn preservation of <u>Porites</u> fragments in the basal lag as compared to the delicate preservation of <u>Porites</u> branches in the upper part of the layer. This pattern cannot be explained hydrodynamically but rather suggests post-event colonisation by the corals.

(2) <u>Post-event colonisation</u>. In several instances, layers of concentrated skeletal material in offbank spillover lobes and in onbank
sand lobes are directly overlain by well preserved branching coral
colonies of <u>Porites</u> in contrast to worn <u>Porites</u> fragments within the
skeletal lags (Fig. 12; 13). Moreover, the very top of some of the
skeletal layers showed more abundant epibenthic pelecypods such as
<u>Lima</u>, <u>Aequipecten</u>, <u>Chlamys</u> and <u>Anomia</u> (Fig. 13). Such a distribution
cannot be explained hydrodynamically, but must have ecological causes.
<u>Porites</u> grows on windward parts of many South Florida carbonate banks
as mostly uncemented branched colonies on skeletal sand and gravel
substrates (Turmel & Swanson, 1976; Ginsburg & James, 1974; Ebanks &
Bubb, 1975; Enos & Perkins, 1977); it is less abundant on muddier
bottoms of bank interior habitats. Generation of onbank sand lobes
during hurricanes modifies the substrate consistency of the bank
interior in transforming seagrass-covered muddy bottoms into sandy or
gravelly shellgrounds. As a biological response, <u>Porites</u> seems to episodically expand its usual bank margin habitat area both onto the bank
flats (onbank sand lobes) and in an offshore direction (offbank
spillovers). Wanless (pers. comm.) has directly observed that <u>Porites</u>
recolonizes such surfaces after storms, but may be later outcompeted
by seagrass.

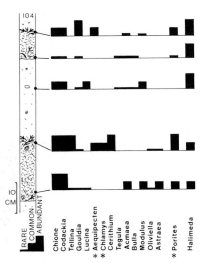

<u>Fig. 13.</u> Faunal composition of bottoms and tops of offbank spillover
layers. New faunal elements in the top parts of these layers
(<u>Aequipecten</u>, <u>Chlamys</u>) or higher abundance (<u>Porites</u>) indicates
post-event colonisation by epibionts (asterix) as a response to the
new substrate provided by the spillover layer.

2.7. Dynamic stratigraphy and history of Safety Valve banks

Based on geomorphological features, Wanless (1969, 1970) has given an overall account on the development of the whole Safety Valve tidal bar belt, in relation to the late Holocene rise of sea level.

FAUNAL CHANGE IN
SHALLOWING-UPWARD SEQUENCES

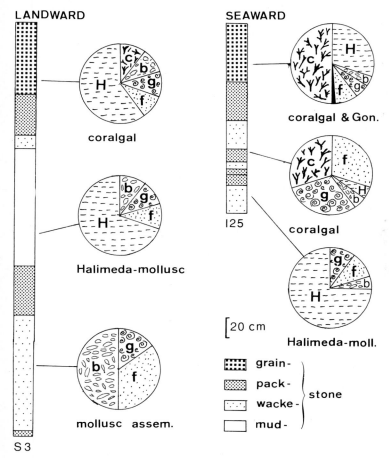

Fig. 14. Two coarsening- and more winnowing-upwards sequences and associated changes in the faunal assemblages that are interpreted as shallowing-upwards trends in the bank history. Cores no. S3 and 125. H = Halimeda, c = corals, b = bivalves, g = gastropods, f = molluscan fragments, black = Goniolithon.

The new core data further substantiate the dynamic history of the banks. A number of cores, especially those taken near the seaward side of the banks or next to tidal channels show "coarsening and more winnowing-upward" sequences, with distinct trends in depositional fabrics and in faunal content (Fig. 14). In core S 3, for instance, the upward transition from wackestone to packstone to grainstone is accompanied by a change from an entirely molluscan to a Halimeda-molluscan and eventually to a coralgal association. Such sequences indicate shallowing-upward trends in the bank history and a transition from a "mudbank" stage into the present situation with distinct skeletal caps, especially at the seaward margins. The seaward zone of coralgal pack- to grainstone is composed of an amalgamation of wedge-shaped skeletal layers (onbank sand lobes) formed during hurricanes. The sand lobe on the present-day bank surface was generated by Hurricane Betsy (1965) and reactivated by later storms such as the January 2o, 1983 storm.

Radiocarbon dating of a double-valved Codackia shell from the basal lag of a skeletal wedge, 90 cm below the sand lobe on the present surface gave an age of 980 $\pm$ 8o years (Fig. 6). This date provides an approximate idea about the possible preservation frequency of major events significant for the bank history: the recurrence interval of individually preserved units is in the order of several 100 years. Thus we are dealing with events rare for human experience, but certainly common on a geological time scale.

2.8. Storm effects in Sandy Key (Florida Bay)

"Sandy Key" represents a "discontinuous marine sand belt" (Ball, 1967) separating Florida Bay in the east from a gently sloping ramp towards the Gulf of Mexico in the west (Fig. 3). Hurricane Donna formed extensive subaqueous and subaerial spillover lobes that are oriented landward (Ball, 1967; see Fig. 15).

Cores from submerged spillover lobes of the Sandy Key sand belt show dm-thick sharp-based fining-up depositional units (Fig. 16D). These are arranged in distinct coarsening-upward sequences (Figs. 15; 16A,B; 17) indicating upward shallowing. In addition, several successive skeletal spillover layers, separated by soil horizons were found in the subaerial part of the Sandy Key island. These observations support Ball's (1967) inference, that the entire sand belt was concentrated episodically by successive increments of spillover lobes deposited during storm flood tides and onshore wind drift. When these tides ebbed, some of the higher parts were eventually exposed and beaches and shell islands could subsequently form.

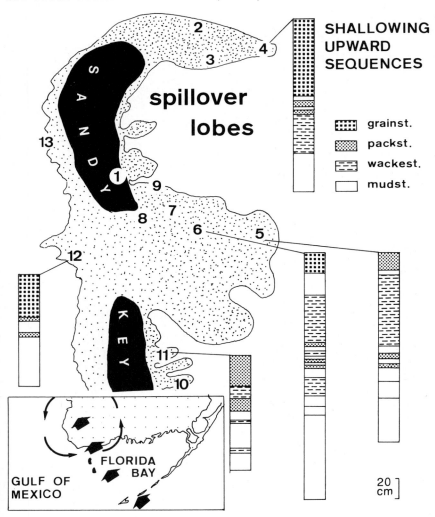

Fig. 15. The Sandy Key area is located at the western edge of Florida Bay. Arrows indicate path and wind circulation of Hurricane Donna according to Perkins & Enos (1968). Hurricane Donna breached the shell island of Sandy Key and generated extensive bayward spillover lobes. Practically all sediment cores taken in this area (numbers 1-13) show shallowing-upwards sequences and record successive increments of hurricane-generated skeletal spillover layers, building up the bank sequence.

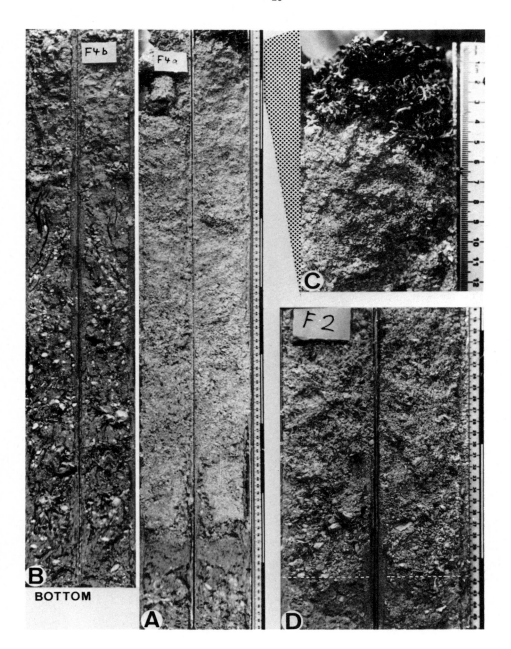

Fig. 16. A) and B) Example of coarsening-upward sequence; note thick cap of skeletal grainstone (spillover lobe) that is colonized by rhodoliths (see close-up of Fig. C); core no. 4 in Fig. 15. D) Skeletal spillover layer with sharp base and well-developed fining-up sequence; core no. 2 in Fig. 15.

COARSENING-UPWARD

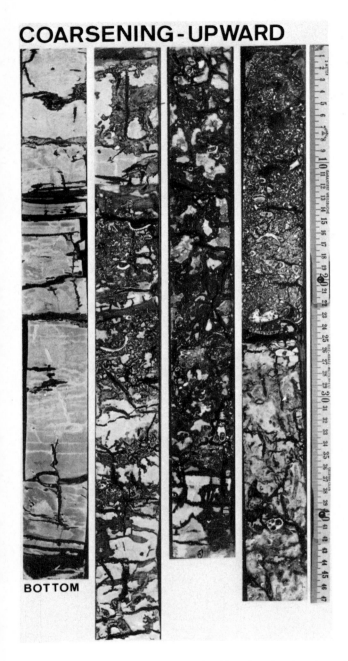

BOTTOM

Fig. 17. Core through Sandy Key bank showing typical coarsening-upward sequence (core no. 5 in Fig. 15). Note upwards transition from layered lime mudstone to thin layers of skeletal packstone (earlier storm spillovers), to bioturbated wackestone to the top spillover unit of skeletal pack- to grainstone produced by Hurricane Donna.

2.9. Summary and conclusions

(1) Storm effects in shallow nearshore water are dominated by onshore directed wind-drift currents. In nearshore skeletal banks such as the Safety Valve (Biscayne Bay) and Sandy Key (Florida Bay) onshore wind stress caused the formation of onshore oriented sediment lobes. These lobes commonly consist of sharp-based, fining-upward layers of skeletal sand and gravel, in contrast to the much larger cross-bedded spillover lobes in Bahamian oolite shoals (Ball, 1967; Hine, 1977). The stratigraphy of the studied banks implies that successive storm-generated lobes contribute significantly to bank accretion and growth. Thus storm effects play an important role in constructing and molding non-reef skeletal buildups in nearshore carbonate environments (see also Wanless, 1978, 1979a).

(2) Three major stages may be distinguished in the storm-affected development of nearshore carbonate banks (Fig. 18):

a) after the initial transgression lag: "mudbank" stage, often localized by bedrock topography (see also Basan, 1973; Turmel & Swanson, 1976). This stage represents the early phase of both the Safety Valve and the Sandy Key banks.

b) Progressive growth into shallower water (high rates of carbonate production) increased the effects and frequency of storms in winnowing and accreting skeletal material as subaqeous sand lobes on the seaward bank margins. These processes cause the differentiation into (1) a bank interior muddy facies, and (2) a windward margin skeletal facies. This stage is represented by the present Safety Valve. Shallowing-up sequences, a common pattern in many types of carbonate banks of modern (e.g. Basan, 1973; Turmel & Swanson, 1976) and ancient environments (e.g. Wilson, 1975) become thus understandable in the context of upward increasing storm impact.

c) Further increments of hurricane-generated sand lobes cause further shallowing and lead eventually to subaerial spillover lobes and to the development of beaches and shell islands. In the fossil record, such stages show evidence for vadose diagenesis, pedogenesis and karstification of shoal tops.

These three stages illustrate the transition from an initial "mudbank" stage into a "shell island" stage. This transition is largely controlled by hurricane spillovers deposited episodically as discrete increments onto the banks.

(3) In conclusion, onshore wind drift and waves during storms and hurricanes contribute to the construction and buildup of nearshore carbonate banks, comparable to the more familiar formation of barrier islands by buildup of submarine shoals. Possible ancient analogues for such storm-molded carbonate banks are nummulite banks in the Eocene of Egypt (Aigner, 1982b; 1983) and crinoidal banks in the Muschelkalk of SW-Germany (see part II).

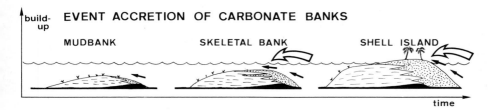

Fig. 18. Model for episodic sediment accretion by storm-induced onshore wind drift as an important factor for the buildup of nearshore carbonate banks. Initial "mudbanks" may develop behind local paleo-highs (e.g. initial Safety Valve behind Key Largo ridge) and become stabilized by seagrass. Onshore directed wind drift during storms piles skeletal material as spillover lobes onto seaward bank margins and with time leads to the development into "skeletal banks" (present stage of Safety Valve). Further increments of storm accretion results in the formation of beaches and "shell islands" (present Sandy Key area).

3. STORM SEDIMENTATION
IN OFFSHORE SHELF AREAS
GERMAN BAY (NORTH SEA)

3.1. Introduction

The Florida example dealt with storm processes in shallow nearshore environments, where onshore directed wind drift currents and waves play a dominant role. In contrast, the effects of storms in offshore subtidal settings involve (1) stirring of the seafloor by oscillatory currents, caused by a deepening of the wave base, and (2) lateral sediment transport from coastal to shelf regions by offshore flowing gradient currents. These gradient currents compensate for wind-stress and nearshore water set-up and are often enhanced by ebb-tidal currents.

Apart from a variety of factors concerning the nature of storms, especially their duration (Allen, 1982), the sedimentary record of storms in the offshore should be largely controlled by two factors: (a) distance from land: the nearshore sand facies serves as a sediment source for offshore storm sand sheets; (b) local waterdepth: storm effects decrease with increasing depth.

The object of this chapter is (1) to provide an actualistic example of storm sedimentation associated with gradient currents in more offshore environments, (2) to demonstrate systematic changes in the nature of storm layers along coastal-offshore transects, and (3) to show how such "proximality trends" may be applied to the analysis of ancient storm-depositional systems.

This chapter is based on work at the Senckenberg-Institute in Wilhelmshaven, supervised by Prof.Dr. H.-E. Reineck, and draws largely from Aigner & Reineck (1982).

3.2. Methods

26 vibrocores, varying in length between 1 and 3 m, and 13 box cores collected along west-east offshore-shoreface transects across the German Bay (Fig. 19) were used for this study. Cores were studied using detailed logging, epoxy peels (relief casts), by grain size analysis (sieving) and partly by X-radiography. For statistical treatment, the following parameters were determined:

1. percentage of sand layers per core;
2. frequency of sand layers (= storm events) per meter core;
3. percentage of wave-ripple cross-lamination against total amount of sand present;
4. quantitative composition of shelly fauna in some storm layers;
5. thickness distribution of storm sands in each core;
6. mean and maximum thickness of storm sand layers;
7. bedform sequences within storm sand layers;
8. degree of bioturbation (using the classification of Reineck, 1963).

3.3. Study area and previous work

The study area is situated in the inner part of the German Bay, between Büsum and the island of Helgoland (Fig. 19). Water depths range between 3m and more than 30 m. This part of the German coast belongs to the macrotidal type with open tidal flats; mean tidal range is 3 m, tides are bidiurnal. Tidal currents trend mainly WNW-ESE and

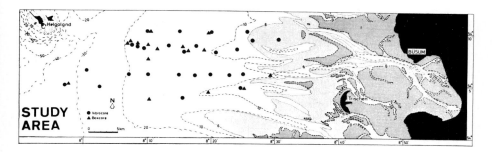

Fig. 19. Study area in the German Bay, North Sea, with positions of vibrocores and boxcores taken (from Aigner & Reineck, 1982).

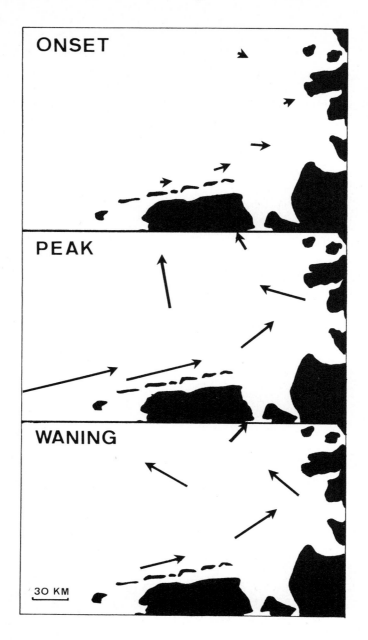

Fig. 20. Current measurements in the German Bay during a storm surge, after Gienapp & Tomczak (1968). Note weak onshore currents during onset of the storm period, in contrast to strong offshore gradient currents during the peak and waning stages of the storm.

exceed 80 cm/sec. Wave attack of the N-S trending shoreline predominates from W and NW, due to strong winds and storms which mainly
blow from this direction. In the eastern North Sea, wave heights of 6
m are most frequent in January-May and in October-December (DHI 1975).
During storm surges, offshore flowing gradient currents, which
compensate for coastal wind set-up, reached 151 cm/sec at 2 m above
the sea bed (Gienapp & Tomczak, 1968; Gienapp, 1973; see Fig. 20).

The sedimentary facies was extensively studied by Reineck et al.
(1967, 1968), Gadow & Reineck (1969), Reineck & Singh (1972) and
reviewed by Reineck & Singh (1980, p. 394-397); consequently detailed
descriptions can be omitted here. In beach-shelf profiles three main
facies belts running parallel to shore have been distinguished: (1)
coastal sand, (2) transition zone, and (3) shelf mud. This particular
part of the German Bay, where a Holocene sediment wedge overlaps
Pleistocene relic sands (Reineck, 1969) may therefore be viewed as an
example of a modern "graded shelf" (cf. Johnson, 1978) in an advanced
stage of equilibrium.

3.4. Storm stratigraphy: descriptions (Fig. 21)

3.4.1. Supra- and intertidal storm layers

Thin, irregular and discontinuous shell layers in supratidal sediments
along the North Sea coast have long been recognized to be caused by
storm flooding (Richter, 1929; Häntzschel, 1936; Reineck, 1962). They
are often lobate or fan-shaped, thus resembling on a much more modest
scale the spillover lobes that have been described in the Florida
example.

On the intertidal flats, Reineck (1962, 1977) and Wunderlich (1979)
have recorded several types of storm deposits. Most commonly they
consist of a sharp erosional base, followed by reworked shells (often
graded), parallel-laminated sands and a wave-rippled top. Reineck
(1977) was able to make a case for in-situ reworking showing that the
depth of tidal flat erosion was equal to the thickness of a new storm
layer.

3.4.2. Shoreface storm layers

Vibrocores from the coastal sand facies commonly show characteristic
sequences ranging in thickness between 5 and 13o cm. The erosive bases
of such sequences are often paved by shell layers, followed by
laminated sand which may be overlain by wave-ripple lamination. Bioturbation is minimal in most cores, but the upper parts of sequences
are often slightly bioturbated. Erosive contacts and amalgamations of
such units are conspicuous (Fig. 21).

Similar sequences have been described by Kumar & Sanders (1976) as
"shoreface storm deposits" from various modern and ancient examples.
They speculated that the geologic record of shoreface sediments
consists largely of storm deposits. This conclusion can be supported
by the present study. Wunderlich (1983) has documented landward
shifting of shoreface shoals and sand ridges in several areas of the
German Bay during storms.

3.4.3. Proximal storm layers

The transition zone of Reineck et al.(1967, 1968) is characterized by
relatively thick (several cm to a few dm) sand layers, which may be
called "proximal" storm sands due to their close proximity to the
shoreface source area (Fig. 21). However, such "proximal" beds may
occasionally also be found in deeper water of shelf mud; these
probably represent exceptionally strong or long lasting storms.

The basal surfaces of proximal storm sands are erosional, often irr-
regularly scoured resembling the guttered surfaces of ancient
analogues. Impact marks may also be observed. These surfaces are often
paved by shell layers, ranging in thickness up to several cm.
Internally, proximal tempestites mostly consist of parallel laminated
sands which are weakly graded. Laminations are often slightly inclined
(low-angle lamination) with minor internal discordances. This type of
stratification may well be equivalent to "hummocky cross-stratification" (e.g. Harms et al., 1975; Hamblin & Walker, 1979; Bourgeois,
1980). Due to the small sample size in vibrocores relative to the
scale of hummocky cross-stratification, however, a definite statement
is not possible. Some of the proximal beds may be entirely composed of
wave-ripple lamination, which is most common at the top of many beds.

STORM STRATIFICATION
cores

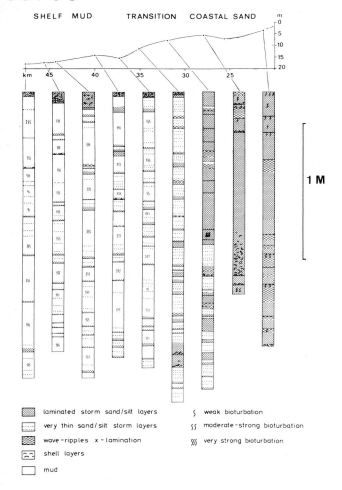

SHELF MUD   TRANSITION   COASTAL SAND

1 M

laminated storm sand/silt layers   ∫ weak bioturbation

very thin sand/silt storm layers   ∫∫ moderate-strong bioturbation

wave-ripples x-lamination   ∫∫∫ very strong bioturbation

shell layers

mud

sequences

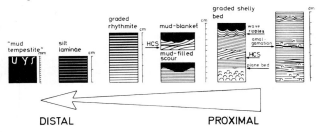

DISTAL   PROXIMAL

Fig. 21. Storm stratification in the German Bay. Upper part: logs of vibrocores collected along one of the nearshore-offshore transects from the coastal sand facies to the transition zone to the zone of shelf mud with decreasing number and thickness of storm sand layers. Lower part: schematic diagram showing proximal and distal facies of typical storm sand sequences(from Aigner & Reineck, 1982).

The top surface may also show mud-filled scours, which so far have only been documented in ancient examples (Goldring & Aigner, 1982). Most of the proximal beds are overlain by a thin unit of non-bioturbated mud which was deposited during the waning stages of the storm events.

3.4.4. Distal storm layers

Thinner (up to just a few mm) and mostly finer-grained sand/silt layers, which predominate in the zone of shelf mud, may be referred to as "distal" tempestites (Fig. 21). They are the lateral deeper water equivalents of proximal tempestites, further away from the coastal sand source. However, such thin sand and silt layers are also very abundant in shallower waters of the transition zone. Here, they document the effects of minor storms which were not strong enough to leave any record in deeper water. The bases of distal storm layers are mostly erosional, although in some cases they may appear non-erosional. Internally, distal storm layers are mostly laminated, sometimes as graded rhythmites, and ripple cross-lamination is conspicuously rare. As in proximal equivalents, the sandy part is typically overlain by a thin blanket of non-bioturbated mud.

Sequences of non-bioturbated mud that sharply cut through underlying burrows and show only minor bioturbation towards the top are interpreted as pure "mud-tempestites". They seem to represent an extremely distal end member of storm sedimentation.

3.5. Proximality trends: results and discussion of statistical treatment

The systematic changes in the nature of storm stratification from coastal sand to proximal and distal ("proximality trends") are documented by means of transects (Fig. 22A) and by maps (Fig. 23).

3.5.1. Percentage of sand

The percentage of sand per meter core decreases consistently with distance from land, while the proportion of mud progressively increases (Fig. 22A, 23). This pattern may be called "graded shelf" (Johnson, 1978), which is due to the decreasing capacity of storm-induced flows to transport sands towards deeper offshore waters.

3.5.2. Frequency of storm layers

The frequency of storm layers per meter core shows an "optimum" in a zone that is roughly equivalent to the transition facies (Fig. 22A,23). Landward, in the coastal sand facies, the frequency of pre-served storm layers decreases, which seems to be due to two factors:

a) in very shallow water, storm layers tend to be highly amalgamated and most probably even "cannibalistic", that is, many storm beds are likely to be wiped out and to be reworked into subsequent ones, which can no longer be singled out;

b) in the shoreface area, storm sequences tend to be several deci-meters thick. Therefore the frequency of storm beds expressed as "per meter core" becomes artificially reduced.

Towards the distal, offshore areas, the frequency of storm layers per meter core decreases continually. Two main factors appear responsible: a) distance from land: the probability of sediment influx from near-shore sand sources decreases away from the shore; b) water depth: since wave effects decrease with depth, erosion and suspension of sediment should decrease in deeper waters.

The "frequency optimum" of preserved storm layers may thus be ex-plained as follows. In water shallower than the "optimum", more storms affect the sea bottom, yet at the same time each tends to erase previous records due to reworking and therefore has a low preservation potential. In water deeper than the "optimum", the preservation poten-tial of storm events is much better and the record more complete (disregarding the effects of post-event bioturbation). Nevertheless, further away from the coast and in deeper water, fewer and fewer storms leave their record on the sea bottom and their transport

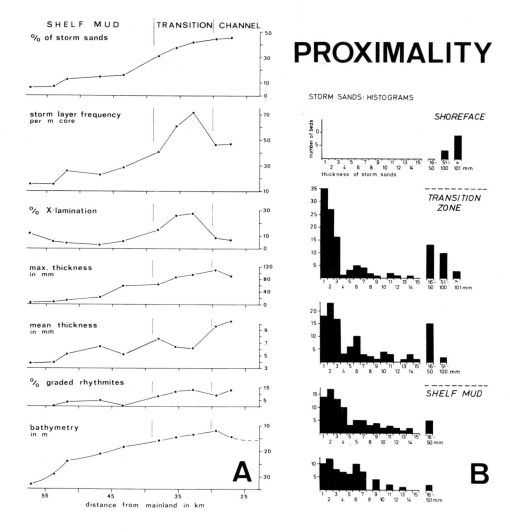

Fig. 22. A) Proximality trends as observed in vibrocores along near-shore-offshore transect. The percentage and frequency of storm sands, the percentage of wave-ripple cross-lamination and of graded rhythmites, the maximum and mean thickness of storm layers decrease with increasing water depth and increasing distance from land. Note also "optimum" of preserved storm layer frequency in transition zone. B) Histograms showing the thickness of storm sands in selected vibrocores along a nearshore (top) - offshore (bottom) transect. In the coastal sand facies (top) only few very thick storm layers are present (shoreface storm deposits). In the transition zone, thicker ("proximal") beds occur in addition to a vast number of thin and very thin ones, whereas in the zone of shelf mud, thin ("distal") storm layers are dominating, but they progressively decrease in number (from Aigner & Reineck, 1982).

PROXIMALITY: MAPS

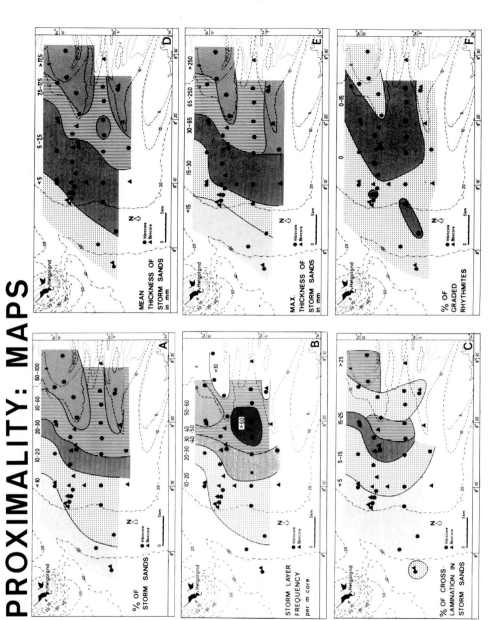

Fig. 23. Proximality trends shown in maps (from Aigner & Reineck, 1982).

capacity decreases, resulting in a decreasing number of storm layers per unit time, and if sedimentation rates are comparable, by unit of thickness.

Within the transition zone, the zone of maximum frequency of storm layers is found in an area just outside the mouth of a major tidal channel (Süderpiep). Similarly, frequency contours are not strictly parallel to the coast but show tongue-like progradation outside the mouth of this channel (Fig. 23). This would suggest, that during storms, sand is mostly transported through tidal channels and spread out on the shelf in a fan-like manner.

3.5.3. Percentage of cross-lamination

From the sand tongues towards the offshore shelf mud, the percentage of wave ripples and wave ripple cross-lamination against the total amount of sand shows basically the same pattern as the frequency of storm layers: first an increase in abundance from very shallow shoreface areas to slightly deeper water (transition zone), then a general decrease towards distal areas (Fig. 22A, 23).

In very shallow water, wave ripples are most commonly formed but they are less frequently preserved because the upper wave-rippled part of the storm sequences is commonly "eaten up" by subsequent storm events. On the other hand, in relatively deeper water, below storm wave base, the absence of ripples may primarily be due to the absence of oscillatory flows.

3.5.4. Storm layer thickness

The thickness distribution of storm layers varies considerably at any one locality, most probably due to the unequal strength of storms over a given area. In general, however, the following patterns are found to be characteristic in histograms (Fig. 22B): (1) in the coastal sand facies, a relatively small number of thick to very thick sand layers ("shoreface storm sands") are present. (2) In the transition zone, there are still abundant thicker beds ("proximal tempestites") in addition to a vast number of thin and very thin storm layers. This may indicate the relatively complete record of the smaller storms which failed to leave any record in deeper waters. (3) The zone of shelf mud exhibits only a small number of storm layers which are generally thin and fine-grained ("distal tempestites"); thicker beds are practically lacking here.

Both the mean thicknesses and the maximal thicknesses of storm layers per core show a general decrease from nearshore to offshore (Fig. 22A,23). A map given by Gadow & Reineck (1969) shows basically the same pattern in the mean thickness of storm sand layers.

3.5.5. Allochthony in storm shell beds (Fig. 24)

Dörjes (in Reineck et al., 1968) has established a zoological zonation of the benthic macrofauna in the German Bay. This well-established zonation allows the use of reworked shells in storm layers as "tracers" of sediment transport during storms.

ALLOCHTHONY IN STORM SHELL BED

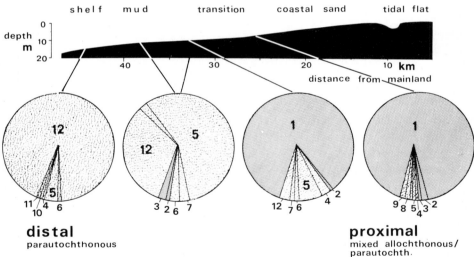

Fig. 24. Reworked shells in storm layers can be used as "tracers" for storm transport. Proximal storm layers are dominated by allochthonous elements imported from tidal flats and tidal channels by offshore bottom return currents (gradient currents). The amount of allochthonous elements decreases away from the shore and distal storm layers are entirely composed of winnowed but parautochthonous soft-bottom assemblages (1 = Hydrobia ulvae, 2 = Cardium edule, 3 = Barnea candida, 4 = Spisula subtruncata, 5 = Nucula nitida, 6 = Abra alba, 7 = Abra nitida, 8 = Macoma baltica, 9 = Angulus tenuis, 10 = Venus fasciata, 11 = gastropods, 12 = Aloides gibba).

Proximal storm beds include a mixture of winnowed parautochthonous bivalves together with allochthonous elements. The latter include Barnea candida, Petricola pholadiformis, Cerastoderma edule, Littorina littorea, Mytilus edulis and Hydrobia ulvae which are clearly introduced from tidal channels or tidal flats respectively. Samples of several storm layers across nearshore-offshore tansects show a consistent decrease in the percentage of allochthonous, laterally introduced shells toward the offshore (Fig. 24). In distal storm layers, shells are largely parautochthonous, winnowed, but generally non-displaced soft-bottom faunas with only very occasional imported forms. These data confirm the view that lateral sediment transport during storms dereases away from the shore.

3.6. Dynamic stratigraphy: storm processes

Several mechanisms have been proposed for sediment supply to the offshore during storm events. Among them are (1) wind-drift currents (Creager & Sternberg, 1972; Morton, 1981), (2) rip currents (Cook & Gorsline, 1972), (3) combined storm-wave density current mechanisms (Hayes, 1967; Hamblin & Walker, 1979; Walker, 1980b; Wright & Walker, 1981), (4) combinations of wave-induced currents with storm ebb currents (Reineck et al., 1967, 1968; Gadow & Reineck, 1969), (5) storm surge ebb currents (Johnson, 1978; Brenchley et al., 1979), (6) purely wave-induced currents (Howard & Reineck, 1981), (7) storm-associated bottom currents combined with storm-wave liquefaction (Nelson, 1982), and (8) combined geostrophic-oscillatory flows (Swift et al., 1983). Allen (1982) presented a detailed semiquantitative model for a combination of wind-induced currents with wave-stirring to explain offshore sediment transport, which is partly applicable to the present situation. Due to the strong tidal influence, however, interaction and enhancement through ebb currents appears to be significant in this case (Reineck et al., 1967,1968; Gadow & Reineck, 1969).

In a way, the German Bay may be considered a "natural laboratory" for storm sedimentation: winds and storms mainly come from the W and NW, i.e. attack the N-S running coastline at a high angle. According to Allen's (1982) model, onshore blowing winds generate drift currents acting in the same direction; these onshore flows are recorded in minor spillover-like supratidal shell layers (Fig. 25) and in onshore sand transport within the shallow shoreface zone as documented by

43

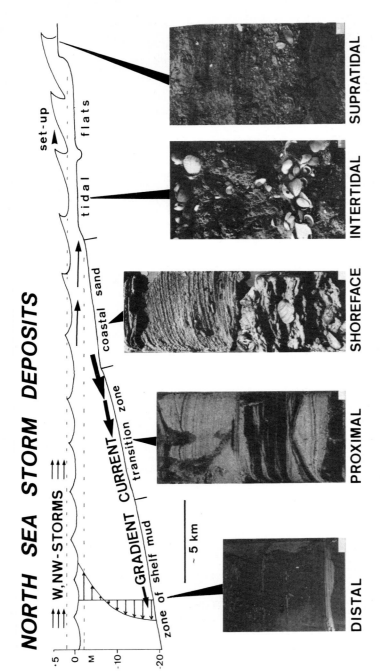

Fig. 25. Summary of storm depositional processes and products in the German Bay, North Sea. Storms from the W and NW induce onshore wind drift currents that cause strong reworking of coastal sands (shoreface storm deposits), erosion on tidal flats (intertidal storm deposits) and minor storm spills onto supratidal flats. Nearshore water set-up is compensated by offshore directed bottom return flows (gradient currents), transporting sands offshore and depositing them as proximal to distal tempestites. Each scale bar = 1 cm.

Wunderlich (1983). Wave stirring causes erosion on tidal flats as
documented by shells being reworked into intertidal storm layers (Fig.
25). On the open shelf and within main channels, however, onshore
winds create gradient currents near the sea bottom, flowing offshore
against the wind-drift and compensating for coastal water set-up.
Gienapp & Tomczak (1968) and Gienapp (1973) have measured such
gradient currents and found velocities as high as 151 cm/sec with west
to northwest flow direction. These measurements confirm (1) the model
of offshore flowing gradient currents, and (2) that gradient current
velocities are sufficiently strong to transport sand and to deposit it
as proximal to distal storm layers (Fig. 25).

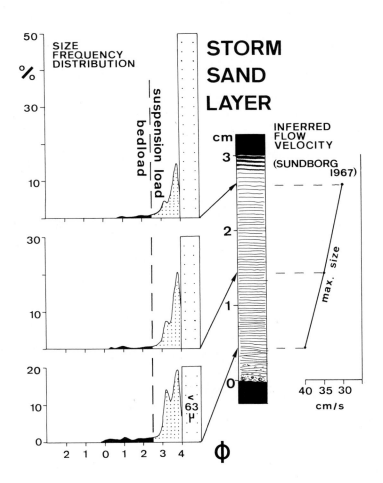

Fig. 26. Grain size analysis (left) and hydraulic interpretation
(right) of one storm sand layer (middle) from the German Bay.

The mode of storm transport can be estimated from the grain size of storm layers. Most of them are fine to very fine sands to silts. Bagnold (1966) and McCave (1971) have shown that quartz grains finer than 17o-2oo μm tend to go directly into suspension (see also Wanless, 1981). Since the storm layers analyzed here are largely finer than 200 μm (Fig. 26), most of the storm sands appear to have been transported in suspension, supporting earlier assumptions of Reineck & Singh (1972).

3.7. Applications (Fig. 27)

The proximality trends documented above by means of transects and maps reflect the decreasing effects of storms with increasing water depth and distance from shore. Similar trends recognized in fossil counterparts should contribute to a better understanding of lateral and vertical facies sequences and should therefore provide a tool for the analysis of ancient storm depositional systems (cf. Aigner, 1982a). The statistical approach has been preferred in order to "even out" the variability of storm parameters (a heavy storm may show the same sedimentary record in offshore areas as a weaker storm nearshore).

In lateral facies sequences, preferentially bound by isochronic surfaces, proximality trends may indicate (1) the source of storm sand, (2) paleobathymetric trends, and (3) basin geometry. Orientation of ripples, scours and impact marks may give further insight into the direction of wave attack and of bottom currents associated with storm events. As suggested by Bourgeois (1980), Hunter & Clifton (1982) and others, water depth and other hydraulic parameters can be estimated by wave ripple characteristics on the top of storm sand layers. Furthermore, proximality trends may be transformed into contour-type maps, which should shed light on two factors: (1) the nature of the sand source: linear sand sources should be reflected in contours parallel to the shoreline and parallel to the bathymetry, in contrast to

Fig. 27. Application of tempestite sequences to ancient storm depositional systems. Proximality trends help to understand lateral and vertical facies sequences and indicate paleobathymetric trends. The sequential and geometrical arrangement of shoreface, proximal and distal tempestite facies may be used in basin analysis as a monitor of relative sea level fluctuations. ⟶

APPLICATIONS
a) facies analysis

PROXIMALITY TRENDS

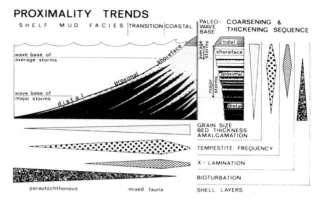

b) basin analysis

MONITOR OF SEA LEVEL CHANGES ?

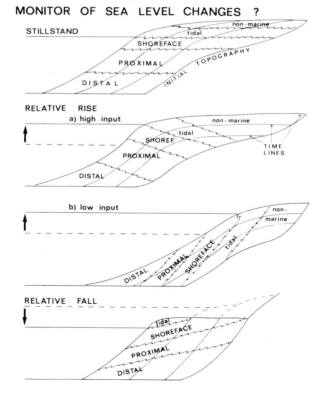

channeled point sources, documented by fan-like progradations of contours; (2) Comparisons of such contour-maps with isopach patterns and paleocurrent data appears to be a potential tool for reconstructing size, shape, geometry and morphology of shallow-marine storm-depositional basins.

Proximality trends should also be valuable in the analysis of <u>vertical facies sequences</u>. Regressive, progradational sequences are widely familiar. They are characterized by an upward coarsening and thickening of storm layers: distal tempestites are overlain by proximal ones, followed by shoreface storm deposits, which in turn may be followed by tidal deposits (Fig. 27). The thickness of the shoreface facies in progradational sequences is - under certain circumstances - the direct response to the wave energy of the coast (Howard & Reineck, 1981). Similar to the method of Klein (1971) to determine paleotidal range, the paleodepth range of storm waves may be determined in such sequences. The boundary between shoreface and transition facies roughly marks the depth of average storm waves (cf. Reineck & Singh, 1980, Fig. 583) while the last occurrence of wave ripples denote approximately the maximal reach of wave-induced oscillatory currents, i.e. wave base of major storms (Fig. 27). Thick transgressive, upward fining and thinning sequences appear to be less common in the geologic record (Bourgeois, 1980).

Proximality trends should be particularly useful in the interpretation of facies cycles in the order of several meters to a few tens of meters in thickness (both fining- and coarsening-upward sequences). It may be speculated, that tempestite sequences may be used as a sensitive monitor of relative sea level changes: the sequential and geometrical relationships of several such sequences with shoreface, proximal and distal tempestites should shed light on small-scale cyclicity in the stratigraphic record caused by relative sealevel fluctuations (Fig. 27). For such analysis, principles of seismic stratigraphy (Vail et al., 1977) have to be applied and isochronic time lines (provided either by lithostratigraphic markers or by index fossils in surface exposures) should be available. These predictions remain to be tested by further field studies in ancient tempestite sequences.

3.8. Summary and conclusions

1. Storm effects in offshore shelf areas such as the German Bay mainly involve (1) the transport of sands and shells from coastal sand sources to the offshore by seaward flowing gradient currents, that compensate for nearshore water set-up, and (2) the deposition of sediment in the form of offshore storm sand sheets (tempestites) intercalated with the background muds.

2. Similar to the Bouma-sequence in turbidites, the following idealized vertical succession with 4 divisions has been found in the North Sea tempestites: sharp erosional base , succeeded by (A) a layer with reworked shells, followed by (B) parallel- or low-angle lamination with internal discordances (probably HCS), overlain by (C) wave-ripple lamination and wave ripples which are topped by (D) a mud blanket. Comparable sequences have been recognized by various authors in both modern and ancient storm depositional systems.

3. Laterally, storm stratification varies significantly due to changes in water-depth and in distance from land. "Shoreface" storm beds are several decimeters thick and mostly parallel or low-angle laminated sands with few shells at the base. Because of a high degree of amal- gamation, the upper wave-rippled division is commonly lacking, while mud blankets are completely absent. "Proximal tempestites" are up to several cm thick and often show "complete" vertical sequences (see 1.). Proximal shell beds consist of mixed faunas including allo- chthonous elements imported from tidal flats and tidal channels. "Distal tempestites" are mm-thin laminated sands and silts without wave-ripples. Shell layers are dominated by winnowed parautochthonous soft-bottom faunas. Mud tempestites can be regarded as extreme end members.

4. Systematic qualitative and quantitative changes in the nature of storm layers ("proximality trends") have been recognized in offshore- nearshore profiles. These reflect the decreasing effects of storm events with (1) increasing distance from the coastal sand source and with (2) increasing water depth. The percentage of sand, storm layer thickness and grain size as well as the degree of amalgamation show a continuously decreasing trend from shallow to deeper water, while bio- turbation generally increases. The frequency of storm layers and the percentage of wave ripple cross-lamination shows an "optimum" in the

transition zone (around 7-15 m water depth) and adjacent to the mouth of a major tidal channel. From there in an offshore direction, these parameters follow the general decreasing trend. Isoline patterns in maps of several of these parameters suggest that during storms, sand is derived from the coast, transported through tidal channels and is finally spread over the shelf bottom in a fan-like manner. The decrease in lateral sediment transport during storms from nearshore to offshore is also documented by the faunal composition of shelly tempestites.

5. Turbidites have been successfully used to reconstruct deep basins in a quantitative way of analysis (Walker, 1967; 1970; Lovell, 1970), although the submarine fan concept has proven more valuable. Despite many differences, similar methods of approach should be applicable for tempestites and for reconstructions of shallow-marine basins. It is therefore suggested that the general "proximality trends" recognized here in modern shelf deposits should prove a useful tool in basin analysis of ancient storm depositional systems. In lateral facies sequences, proximality trends should indicate (1) the source area of storm sands and (2) paleobathymetric trends. Maps computed from proximality parameters should help to reconstruct the geometry and facies organisation of shallow-marine basins. In vertical progradational sequences it may be possible to estimate the depth range of paleo-storm waves under ideal circumstances. The sequential and geometrical relationships of shoreface, proximal and distal tempestite facies may even be used as a monitor of relative sea level changes and thus contribute to a better understanding of cyclicity in the stratigraphic record (Fig. 27).

4. FINAL REMARKS

ON

ACTUALISTIC MODELS

The principle of uniformitarianism, "the present is the key to the past" is one of the most widely used concepts in sedimentary geology. In recent years, however, limitations to the uniformitarian approach are being increasingly realized and discussed (Reineck & Singh, 1980; Ager, 1981; Shea, 1982; Dott, 1983). In particular, it is now more generally accepted that rare, catastrophic events are extremely significant in the stratigraphic record; Ager (1981) consequently suggested the term "catastrophic uniformitarianism".

A particular problem are actualistic models of shelf sedimentation and their application to epicontinental sequences because "we simply have no existing models for epeiric seas" (Irwin, 1965). In very recent years, however, more data on modern seas that may be considered somewhat analogous to ancient epeiric seas could be gathered, especially from the Bering Shelf (Nelson & Nio, 1982) and the North Sea (Nio et al. 1981).

Since general processes and patterns of sediment dynamics are likely to be more strictly uniformitarian, it seems justified to use sedimentary processes and products from modern "epeiric-like" seas as a guide to interpret ancient epicontinental deposits. In the following second part of this study, elements of the above actualistic models are applied to the epicontinental M. Triassic Muschelkalk Basin. It is shown that some aspects of modern Floridan skeletal banks may be analogous to shallow-water skeletal banks and that the modern German Bay (North Sea) shows parallels to deeper-water settings of the Upper Muschelkalk. Modifications arise, however, from influences of the Coriolis effect.

Part II

AN ANCIENT STORM DEPOSITIONAL

SYSTEM:

DYNAMIC STRATIGRAPHY

OF INTRACRATONIC CARBONATES

UPPER MUSCHELKALK

(MIDDLE TRIASSIC)

SOUTH-GERMAN BASIN

1. INTRODUCTION

1.1. Scope of study

Epicontinental sequences are common throughout the Phanerozoic. Their descriptive stratigraphy is mostly well-established, but the dynamic processes that control such cratonic seas are far from being fully understood. More recently, however, new approaches and techniques and a refined understanding of sedimentary processes have greatly advanced our ability to decipher the stratigraphic record in terms of a process-oriented stratigraphy. This new approach has been summarized by Matthews (1974, 1984) as "dynamic stratigraphy". Similarly, but on a smaller scale, recent work within the Tübingen SFB 53 has focussed on the significance of bedding and stratification (e.g. Einsele & Seilacher, 1982). The object of this chapter is to further develop the theme of "event-stratification" and to incorporate it into a broader approach of "dynamic stratigraphy". This is exemplified with the Triassic Upper Muschelkalk in SW-Germany, a shallow epicontinental basin in which the decriptive stratigraphy is well established.

In contrast to continental-margin "drop-off" shelves, many epi-continental carbonate sequences represent gently inclined "carbonate ramps" (Ahr, 1973). Carbonate ramps are characterized by banded dis-tribution of facies patterns parallel to the coastline with the highest energy deposits near the shore passing downslope into deeper water, low-energy deposits. Ramps differ from rimmed "drop-off" shelves by the absence of a sharp hypsographic break and of continuous reef trends. The lack of such protective reef barriers makes ramps very susceptible to the effects of swell, waves and storms. Consequently ramp systems should be hydrodynamically different from rimmed shelves and open platforms, with storm processes playing a major role.

This chapter integrates stratigraphic, sedimentologic and paleo-ecologic data from the epicontinental carbonate sequence of the Upper Muschelkalk in order to reconstruct the dynamics of stratigraphic accumulation in this type of storm-dominated intracratonic basins.

1.2. Hierarchical approach to dynamic stratigraphy

Moving progressively from smaller to larger scale, three levels of stratigraphic sequences are analyzed in a bed-by-bed manner (Fig. 28):

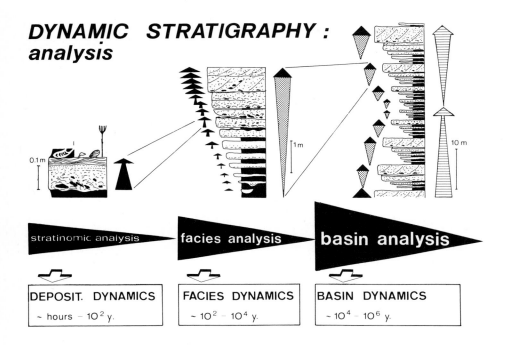

Fig. 28. Schematic diagram illustrating simple approach towards a "dynamic stratigraphy" of storm-dominated basins based on a hierarchical analysis of three levels of stratigraphic sequences (from Aigner, 1984).

1. At the lowest level, individual strata are analyzed as the smallest depositional units and as the basic building elements of stratigraphy. Stratification and facies types, bio- and ichnofabrics allow the re-construction of <u>depositional dynamics</u> (e.g. erosional/depositional processes, mode of transport, substrate changes and colonisation patterns) over short time spans.

2. At an intermediate level, lateral and vertical development and changes of facies (facies sequences) and corresponding changes of biota over longer time scales are analyzed. Such sequential analysis sheds light on <u>facies dynamics</u> (e.g. transgressive/regressive cycles).

3. At a still higher level, the sequential and geometrical arrangement and packaging of facies sequences within the whole depositional basin is analyzed over still longer time intervals. The hierarchy of cycles and their regional distribution patterns give insights into the <u>basin dynamics</u>, i.e. the effects of different scales of baselevel changes (sealevel, subsidence etc.).

1.3. General setting and stratigraphy (Fig. 29)

According to Ziegler (1982), Triassic basins in Western and Central Europe are controlled by regional crustal extension that induced the subsidence of a complex network of grabens and troughs. They reflect global plate reorganisation and early disintegration of the Pangean supercontinent (Fig. 29A). In some of the thus formed intracratonic basins, notably in the Central European German Basin, the Triassic is characterized by the so-called Germanic facies province, i.e. a tri-partite subdivision into (1) continental Buntsandstein clastics, (2) Muschelkalk carbonates and evaporites, and (3) continental Keuper redbeds.

The Germanic Muschelkalk represents a semi-enclosed shallow-marine intracratonic basin, separated by the Vindelician High from the open Tethys sea (Fig. 29B). A relative rise in sealevel in the early Anisian (Kozur, 1974) induced the Lower Muschelkalk transgression and free communication between the Tethys and its marginal basins in Central and Northwestern Europe. Restricted connections with the Tethys during the later Anisian (Kozur, 1974) caused deposition of Middle Muschelkalk evaporites. During the very late Anisian (Kozur, 1974), renewed communication of the German Basin with the Tethys in- duced the Upper Muschelkalk transgression and the reestablishment of open-marine clear-water conditions with carbonate sedimentation. The later Ladinian marks a transition to marginal marine and continental sedimentation. According to Marsaglia & Klein (1983), the paleo- latitude of the German Muschelkalk Basin was within the "hurricane- dominated" and "winter-storm" zone in their paleo-storm model (Fig. 29A).

The Upper Muschelkalk shows the following, very generalized facies provinces from its margin to its center (Fig. 29C): (1) a zone of coastal clastics northwest of the Vindelician High (Schröder, 1967),

(2) a zone of variable width composed of partly dolomitic rocks with lagoonal aspects, (3) a belt of massive, nearshore carbonates including thick accumulations of skeletal and oolitic sediments, (4) alternations of thin argillaceous limestones and marlstones in offshore, basinal areas.

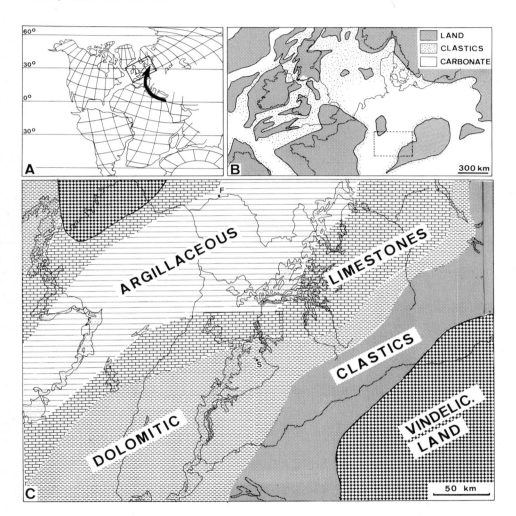

Fig. 29. General setting of the Upper Muschelkalk. A) Plate tectonic reconstruction of the Triassic world (after Smith et al., 1981, for Anisian); arrow: possible pathway of storms and hurricanes using paleo-storm model of Marsaglia & Klein (1983). B) Paleogeography of the Muschelkalk (Anisian and Ladinian) in Central Europe (simplified after Ziegler, 1982). C) Strongly generalized organisation and overall facies provinces in the South-German Basin during the upper part of the Upper Muschelkalk, discoceratites-beds (compilation of data from Wagner, 1913 a,b; Schröder, 1967; Frank, 1937; Schäfer, 1973; Duringer, 1982). Due to missing outcrops, facies boundaries are hypothetic in some areas. S = Stuttgart, F = Frankfurt (from Aigner, 1984).

Descriptive lithostratigraphic subdivision in the Upper Muschelkalk of SW-Germany is based either on thin marlstones or on bioclastic marker beds. Among the large number of marlstone horizons, the more prominent and widespread ones are numbered (e.g. "Tonhoriont alpha, beta, gamma" etc., cf. Figs. 71-76). The bioclastic marker beds are named and comprise a few thin but exceptionally widespread brachiopod shell beds (e.g. "Cycloides-Bank") and a large number of thicker skeletal and oolitic units (e.g. "Trochitenbank 4", "Mittlere Schalentrümmerbank", cf. Figs. 71-76). The marlstone horizons are most prominent in off-shore areas, and many wedge out shorewards. In contrast, the skeletal and oolitic units are best developed in nearshore areas but become less distinct offshore (Fig. 78). (In this study, the term "marlstone" is used for mostly gray, blue or yellowish poorly indurated argillac-eous calcilutite with about 20 - 40 % clay.)

1.4. Previous work

Biostratigraphic zonations are based on ceratites and on conodonts and were summarized by Wenger (1957) and by Kozur (1974) respectively. In SW-Germany, most work has been concerned with lithostratigraphic sub-division and correlation. Following pioneer work of Wagner (1913a,b), Upper Muschelkalk lithostratigraphy was refined mainly by Paul (1936), Vollrath (1938-1970), Wirth (1957), Merki (1961), Geyer & Gwinner (1968), Aust (1969), Brüderlin (1969), Gwinner (1970), Bachmann & Gwinner (1971) and Gwinner & Hinkelbein (1976). Microfacies studies include Brüderlin (1970), Skupin (1970), Bachmann (1973) and Schäfer (1973).

More recently, various paleoecological and sedimentological aspects were studied by Reif (1971, 1982), Aigner (1977-1984), Aigner, Hagdorn & Mundlos (1978), Aigner & Futterer (1978), Hagdorn (1978-1985), Hag-dorn & Mundlos (1982, 1983), Bachmann (1979), and Mehl (1982). In several of these papers the importance of episodic events such as storms has been recognized, emphasizing the concept of "event-strati-fication" (Einsele & Seilacher, 1982). Independantly, French geologists presented facies reconstructions for the western side of the Upper Muschelkalk Basin (Demonfaucon, 1982; Duringer, 1982, 1984).In addition to storms, Duringer (1982, 1984) considered tsunamis to be responsible for sedimentary events.

1.5. Methods

41 quarry sections in the Upper Muschelkalk throughout SW-Germany were logged in detail (Fig. 30). Partly based on existing literature, these logs were correlated and assembled into a series of nearshore-offshore transects through the South-German Basin.

More than 1,000 samples (hand specimens, polished slabs; about 100 thin sections) were analyzed in terms of sedimentary structures, micro-facies, trace fossil content etc. Faunal associations were only briefly surveyed; they will receive a comprehensive treatment by Hagdorn (in prep.).

Paleocurrent data were collected from about 60 localities throughout SW-Germany. Specifically, samples of about 100 beds, marked with their original orientation, were extracted from quarry walls and after cleaning were studied for sole mark orientation.

Selected specimens of fine-grained limestone were studied under the SEM and searched for nanno-organisms.

For a few samples (hardgrounds, underbeds, steinkerns etc.) oxygen and carbon isotope composition was determined. But although circumstantial evidence strongly suggested different origins for these samples, their isotopes showed only negligible variation suggesting that diagenetic processes (recrystallisation) have distorted any original isotopic signal.

Specimens are deposited in the Institute and Museum for Geology and Paleontology, University of Tübingen (no. 1629).

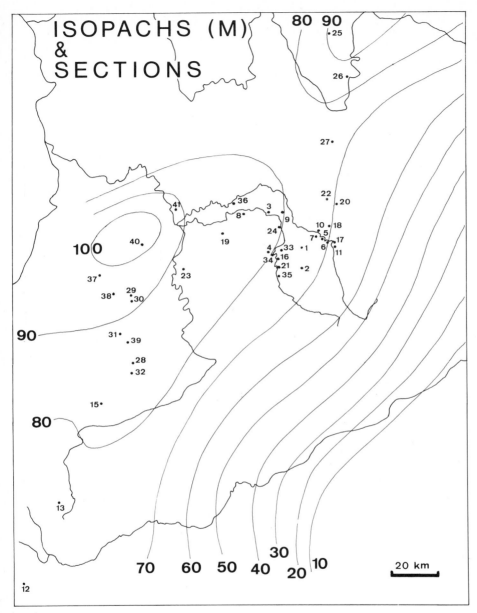

Fig. 30. Location of measured sections in Upper Muschelkalk of SW-Germany; isopachs according to Schäfer (1973). Sections: 1 = Steinbächle, 2 = Ummenhofen, 3 = Garnberg, 4 = Gottwollshausen, 5 = Neidenfels, 6 = Tiefenbach, 7 = Lobenhausen, 8 = Forchtenberg, 9 = Nitzenhausen, 10 = Kirchberg, 11 = Heldenmühle, 12 = Bonndorf-Boll, 13 = Zimmern, 14 = Heming (France), 15 = Mötzingen, 16 = Steinbach, 17 = Barenhaldenmühle, 18 = Brettenfeld, 19 = Weißlensburg, 20 = Gammesfeld, 21 = Wilhelmsglück, 22 = Schmalfelden, 23 = Ilsfeld, 24 = Rüblingen, 25 = Werneck, 26 = Dettelbach, 27 = Aub, 28 = Darmsheim, 29 = Illingen, 3o = Roßwag, 31 = Heimsheim, 32 = Ehningen, 33 = Eltershofen, 34 = Heimbach, 35 = Westheim, 36 = Berlichingen, 37=Bretten, 38 = Ötisheim, 39 = Malmsheim, 40 = Gemmingen, 41 = Gundelsheim.

2 . STRATIFICATION

AND FACIES TYPES

2.1. General

Bedding in sedimentary rocks has long attracted geologists and strati-
graphers. Early work on stratification has focussed on the description
and classification of stratification types (e.g. McKee & Weir, 1953;
Kelly, 1956; Shrock, 1948), although some attempts in interpreting the
nature of stratification have been made (e.g. Andree, 1916; Brinkmann,
1930).

More recently, studies in "comparative sedimentology" have attempted
to understand the physical, biological and diagenetic processes, that
are responsible for bedded sequences (e.g. Harms et al., 1975; Keary &
Keegan, 1975; Einsele & Seilacher, 1982). Studies in modern environments indicate that distinctive types of layering may be significant
for certain environments and sedimentation events (e.g. Gebelein,
1977; Hardie & Ginsburg, 1977; Reineck & Singh, 1980). It therefore
seems desirable and promising to study individual beds and strata in
great detail in order to understand the dynamic processes that
generate stratification. Such an approach, centering around the facies
of individual beds, i.e. all physical, biological and diagenetic
attributes of single strata, might be called "stratinomic approach";
according to Lombard (1978) "stratinomy deals with the accumulation
and succession of layered sequences...related to the dynamics of the
depositional environment".

Such a "stratinomic approach" has the following advantages:

1. Bed-by-bed analysis depicts the smallest rock- and time-stratigraphic units (often identical to single sedimentation events) that
represent both the basic building element of every sedimentary facies
and the shortest time spans represented within the stratigraphic
record.

2. Stratification types are easily recognizable, not only in surface
exposures, but also in subsurface cores.

3. A refined understanding of single beds as "facies elements" helps to refine more general "facies models" and paleoenvironmental inter- pretations.

4. Stratification sequences and microstratigraphic sequences can often be used to deduce source, transportation and depositional process, as well as biologic and diagenetic overprints.

5. Similar, if not identical stratification types occur in similar settings throughout the rock record. In analogy to Wilson's (1975) "standard microfacies types", it is desirable to establish "standard stratification types" that serve as a quick guide for depositional interpretations. In a way, standard stratification types fill the gap between "standard microfacies types" (Wilson, 1975) and "standard facies models" (Walker, 1980a).

So far, studies on sediment dynamics have been mainly carried out for clastic rocks (e.g. Harms et al., 1975), while carbonates were largely neglected. The following is an attempt to reconstruct some aspects of sedimentary and ecological dynamics reflected in stratification phenomena of a storm-dominated carbonate sequence. Particularly, the following questions are adressed:

1. How can stratification types be used as environmental indicators ?

2. To what extent can depositional dynamics be reconstructed based on stratification characteristics ?

3. Are there paleoecological patterns related to stratification and to short-term physical processes ?

4. To what extent is primary bedding modified, enhanced or obliterated by diagenetic processes such as pressure solution ?

5. what is the significance of bedding planes ?

2.2. Peritidal strata (Fig. 31)

In the study area, carbonates of the peritidal suite are rather rare and largely restricted to the transitional zone between the Middle and

Upper Muschelkalk. Three main types of stratification can be distinguished:

1. <u>Flat laminations:</u> Finely laminated carbonates, with individual laminae being slightly crenulated (Fig. 31A), domed and sometimes mud-cracked (Fig. 31B) resemble "cryptalgal limestones" described by many authors (e.g. Aitken, 1967; Reinhard & Hardie, 1976; James, 1980). They are interpreted as algal laminites, in analogy to structures in modern inter- and supratidal carbonate environments (Logan et al., 1964; Hardie, 1977; Shinn, 1983).

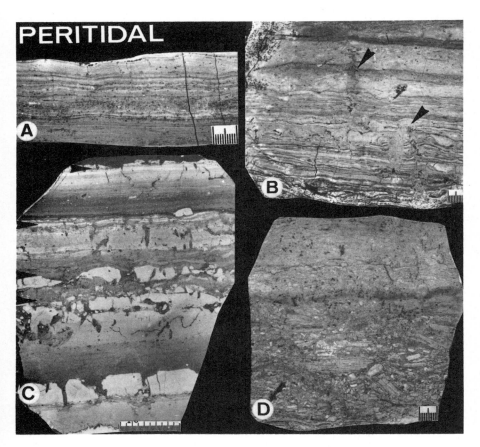

<u>Fig. 31.</u> Peritidal strata. A) Finely laminated carbonate, interpreted as algal laminite (Untertalheim quarry). B) Mudcracks (arrows) in algal laminite (Nennig, Luxembourg). C) Several fining-upward pelmicrite and micrite layers interpreted as supratidal storm layers (Zimmern quarry). D) Reworked clasts of algal laminites in a fining-upward sequence ("flat pebble conglomerate") (Nennig, Luxembourg).

2. Mud-cracked fining-up layers: Cm-thick layers of fining-up micrite and pelmicrite that show at their tops dessication cracks, burrows and thin algal laminites (Fig. 31C). Identical structures, modern and ancient, have been described by Shinn (1983) as lime mud layers deposited by storms and hurricanes onto supratidal flats.

3. Limestone flat pebble conglomerates: Cm- to dm-thick layers with relatively flat, often laminated intraclasts, that may show grading (Fig. 31D). Such layers represent eroded and redeposited algal laminites. In analogy to modern counterparts (e.g. Shinn, 1983), they may represent either supratidal storm layers or basal lag deposits in tidal channels.

2.3. Oncolitic wacke- to packstone

Description. This facies occurs most typically in a relatively narrow belt along the margin of the Upper Muschelkalk Basin (cf. Fig. 60C) and consists of 1/2 - 2 m beds that contain abundant Sphaerocodium kokeni, a Girvanella-oncoid (Wagner, 1913b; Peryt, 1980). Bed geometries are commonly lensoidal and channel-like with higher oncoid concentrations near the base, but sheet-like sediment bodies also occur. Microfacies are dominated by unsorted and muddy packstones and wackestones, sometimes grainstones. Bioclastic grains or intraclasts serve as nuclei for oncolite growth; in fact, most bioclasts show incipient algal coatings or at least thick micritic envelopes. Oncolites vary in size between 5 and 3o mm and are generally irregularly spherical or lensoidal in shape. Larger oncoids may show multi-phase algal coatings, often interrupted by phases of boring activity.

Discussion. According to Wilson (1975), oncolites are typical for shallow, relatively quiet, back-reef (back-bank) environments where they form on the edges of ponds and channels. The typical oncolite facies of the Muschelkalk occurs in a similar position, landward of a belt of oolitic grainstone (back-bank environment). Units with unsorted and muddy wacke- to packstone reflect a relatively low-energy "lagoonal" environment where oncoids are autochthonous or parautochtonous. However, channel structures with reworked oncoids forming basal lag deposits reveal episodic hydrodynamic activity. Multi-phase oncoids also demonstrate that oncoid growth often took place intermittently, interrupted by phases of burial and bioerosion (borings) followed by renewed algal coating. In analogy to modern oncolite facies of the Bahamas (Hine, 1977), storms may be responsible for these reworking episodes.

2.4. Massive oolitic pack- to grainstone

Description. Thick (1-5m) beds of ooid and oolitic pack- to grainstone are most common in the marginal limestone facies province of the Upper Muschelkalk (cf. Fig. 60C), where they commonly occupy distinct, shore-parallel belts, about 10 km in width (see maps of Vollrath, 1955a: Fig. 7). Detailed mapping of Hagdorn (1982) showed that elongated shoal bodies can be distinguished within such belts. The bases of such units are either sharp with erosional relief or show more gradual transitions from underlying skeletal wacke- into pack-stone. Oolitic beds are massive, light-coloured, clean limestones that contrast with the under- and overlying thinner-bedded, darker and more argillaceous carbonates. They consist of well-sorted oosparites and oo-biosparites. Ooid diameters range between 0.3 - 0.8 mm (cf. Bachmann, 1973). Molluscan shell debris is generally worn and rounded, most particles having well-developed micritic envelopes. Faint large-scale cross-stratification becomes evident in weathered sections , with (truncated) sets ranging from a few decimeters to more than a meter in thickness. Paleocurrents are directed mostly longshore (NE), although onshore (E,SE) and offshore (W,NW) directions may also occur. Top surfaces are sharp and some show Fe-rich veneers or rare hardgrounds that are colonized by boring organisms (e.g. Calciroda) and encrusted by the bivalve Placunopsis ostracina, which sometimes forms small biostromal patches (Bachmann, 1979; Hagdorn, 1982).

Discussion. Similar ooid and oolitic grainstones have been described by several authors from modern environments (e.g. Ball, 1967; Loreau & Purser, 1973; Logan et al., 1969; Hagan & Logan, 1974; Hine, 1977) and shore-parallel sand belts deposited along the western margin of the Persian Gulf are probably the best analogues for the present example (Loreau & Purser, 1973). Therefore, massive oolitic grainstones are interpreted as shoals and banks of carbonate sand deposited as mega-ripples and sandwaves in very shallow agitated water.

Using the approach of Heller et al. (1980) to deduce hydrodynamics, paleovelocities of the Muschelkalk ooids (diameter 0.3 - 0.8 mm) range between 40 - 70 cm/sec. These values correspond to velocities of tidal currents in the Persian Gulf (Purser & Seibold, 1973) or the Bahamas (Ball, 1967; Hine, 1977). Although tidal currents are important in maintaining modern ooid shoals, Ball (1967) and Hine (1977) have shown that large hurricanes play a key role in sand body movement and geometry. Hine (1977) found that in the Lily Oolite Bank (Bahamas), tidal currents are insufficient to cause wholesale movement of the entire sandbody under fair-weather conditions, and concluded that shoals migrate and spillover lobes develop only during large storms. Since bankward energy flux is dominant during such events, storm-generated structures in Lily Bank consist of bankward oriented cross-stratification (migrating megaripples, sandwaves, spillover lobes). The landward oriented cross-stratification of Muschelkalk ooid shoals can be interpreted in a similar fashion, while the occasional offshore orientation may be related to offshore spillover lobes. Longshore paleocurrents indicate alongshore migration of shoals, similar to Cambrian oolite shoals in the eastern USA (Sternbach & Friedmann, 1984).

2.5. Massive shelly pack- to grainstone (Fig. 32)

Description. Thick (1-3m) beds of massive shelly pack- to grainstone are closely associated with the oolitic grainstone, in both vertical and lateral direction. Regionally, this facies forms a flanking zone seaward of the oolite facies belt (Fig. 60C); if it occurs on the landward side of this belt, it is generally finer grained and includes higher proportions of black pebbles and coated grains. Bases may either be sharp with erosional scours, sometimes floored by intra-clasts, or they may show more gradual transitions from underlying skeletal wackestones. Top surfaces are similar to those of the massive oolitic grainstones. Massive skeletal beds range from relatively well-sorted grainstones with abundant micritic envelopes to poorly sorted packstones without micritic envelopes. Grain size varies between fine calcarenites and coarse calcirudites.

In weathered sections, the following stratification types can be recognized:

a) only few units display large-scale cross-bedding, suggesting deposition as marine sand waves (Fig.32A);

Fig. 32. Main stratification types in massive shelly limestones. A) Large-scale cross-stratification, interpreted as marine sand wave (Nennig quarry, Luxembourg). B) Superimposed sets of medium-scale cross-stratification, interpreted as migrating megaripples (Bretten quarry). C) Superposition and amalgamation of thin skeletal sheets and wedges (Ummenhofen quarry). D) Close-up of bedding contact between individual skeletal sheets of Fig. C): note stylolites and irregular seam of residual marl as a product of pressure solution. E) Plan view of internal bedding plane with stylolite owing his shape to a seastar "iden". E´) Side view of the same stylolite (camera cap = 5 cm).

b) more common are superimposed sets of medium-scale cross-bedding, both of the planar and festoon types, suggesting deposition as a series of migrating megaripples forming shoals (Fig. 32B);

c) vertical and lateral amalgamation of wedges, lenses, channels and sheet-like geometries to form a composits unit are more common (Fig. 32C);

d) internally more or less homogenous or faintly parallel bedded sediment bodies also occur, suggesting blankets and shallow banks of skeletal sand;

e) thorougly bioturbated packstones with abundant _Teichichnus_-spreiten and remnants of sharp-based fining-upward layers also can be observed. This lithofacies is transitional to nodular wacke- to packstones.

Discussion. Microfacies characteristics and the preferred strati-graphic position seaward of the oolitic grainstone facies belt suggests an environment seaward of, and slightly deeper than the oolite banks. Similar facies types have been described by Anderson (1972), Holloway (1983) and Ruppel & Walker (1982) and are interpreted as shallow skeletal sand banks and shell shoals. Paleocurrent patterns (see below) indicate that these skeletal banks were, similar to the oolite banks, deposited in a shore-parallel belt as a series of along-shore or shoreward oriented sand waves, megaripples, spillover lobes and sand blankets.

A possible modern analogue may be the "marine sand belt" of a modern carbonate ramp on the western edge of Florida Bay (Ball, 1967; see part I), where skeletal sands are concentrated in a belt along a small break in slope. According to Ball (1967) and my own observations (see part I) the entire sand belt was built episodically by discrete incre-ments during storms. A similar episodic concentration and activation in response to storm flows may be assumed for the Muschelkalk shelly banks.

The finer-grained and bioturbated but massive skeletal packstone facies landward of the oolitic facies belt indicates more protected parts of the shoalwater complex, that was only episodically affected by storm reworking (thin fining-up layers).

Bedding planes. Internal bedding planes within the massive skeletal or oolitic facies are commonly affected by, in some cases possibly entirely caused by, pressure solution. In many instances, massive units are subdivided by bed-parallel stylolite seams on which residual clay minerals may be concentrated to form mm-thin marl-partings ("stylo-cumulates", see Fig. 32D). This produces "stylo-bedding" in the terminology of Logan & Semeniuk (1976). Stylo-bedding planes may cut across primary sedimentary bedding (e.g. cross-bedding foresets); such bedding planes are entirely diagenetic in origin. Barrett (1964) described a similar example of post-depositional stratification in bio-clastic calcarenite and estimated that 4-16 % of the original rock was dissolved during the formation of pressure solution seams.

Taphonomic evidence further substantiates the diagenetic nature of such bedding planes: certain fossils on such surfaces behave as resistant "idens" (terminology of Logan & Semeniuk, 1976) surrounded by stylolites. A spectacular example of this are seastars (Trichasteropsis sp.) that coined the shape and outline of stylolites on such stylo-bedding planes (Fig. 32E,E´).

2.6. Massive crinoidal limestone (Fig. 33)

Description. The so-called "Trochitenkalke" are massive, o.5-5 m thick units of crinoid-rich skeletal packstone with some wackestone and rare grainstone. Within the central parts of the Muschelkalk Basin, many of these beds have been named (e.g. "Trochitenbank 1,2," etc.) and used as markers for lithostratigraphic correlation. They are most prominently developed towards the basin margin (e.g. "Marbacher Oolith", "Crailsheimer Riffkalk") where they are often oolitic and where the crinoid content is highest. In these regions, small pele-cypod/crinoid bioherms are associated with the massive crinoidal lime-stones (Hagdorn, 1978).

Since microfacies types of the "Trochitenkalk" have been established and thoroughly documented by Skupin (1970) and Hagdorn (1978), the reader can be referred to these studies. In weathered sections, the following main types of stratification can be recognized:

a) relatively complex patterns of low-angle and medium-scale cross-stratification (Fig. 33A,B), interpreted as migrating bedforms;

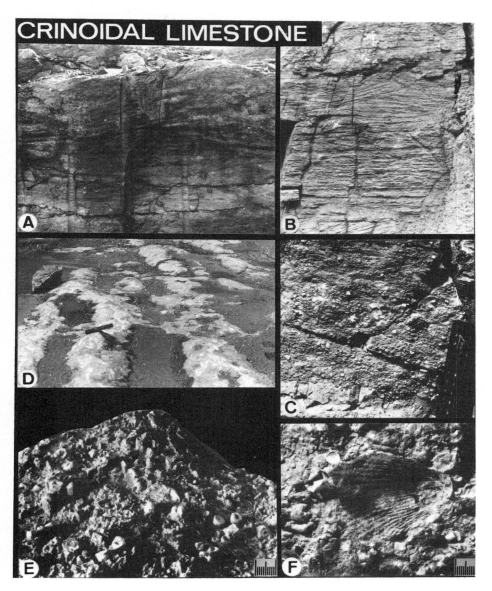

Fig. 33. Stratification types in massive crinoidal limestone. A) and B) Superimposed sets of medium to small-scale cross-stratification and planar lamination; height of both views about 1.5 m (A: Neidelfels quarry, foto B courtesy of H. Hagdorn). C) Sharp-based graded sequence of crinoidal debris. D) Wave-rippled bedding plane colonized by the bivalve <u>Placunopsis</u> <u>ostracina</u> forming small biostromes (Neidenfels quarry). E) - F) Bedding planes affected by pressure solution: E) crinoid ossicles sittig on stylolite "pedestals", F) rip pattern of <u>Lima</u> "telescoped" into bedding plane and into crinoid ossicles (the shell itself is dissolved away).

b) rare cm- or dm-thick layers with sharp bases and fining-up sequences (Fig. 33C), interpreted as storm deposits;

c) remnants of large-scale low-angle stratification consisting of cm-thick crudely graded layers that might represent storm spillovers;

d) in many instances, crinoidal limestones have a massive unstratified appearance; these are probably homogenized by bioturbation.

Massive crinoidal limestones are strongly affected by pressure solution, both in thin-section scale (stylolitic contacts between individual crinoid ossicles) and in outcrop scale (stylolite seams, stylobedding).

Discussion. Massive and commonly ill-sorted crinoidal limestones are known from shallow-water carbonate environments from the Paleozoic to the Jurassic and have been described as "reef-like" or "biostromal" rock bodies, "bioherms", "banks", "mounds", and sandwaves (e.g. Vollrath, 1957, 1958; Harbough, 1957; Carozzi & Soderman, 1962; Cain, 1968; Jenkyns, 1971; Ruhrmann, 1971; Hagdorn, 1978; Brett, 1983).

The massive crinoidal limestones in the marginal parts of the Muschelkalk Basin were interpreted by Linck (1965) as accumulation of allochthonous biodetritus. Hagdorn (1978), however, presented convincing paleoecological evidence that many of these massive accumulations are largely parautochthonous, while some were deposited on paleohighs in very shallow agitated water as subtidal bars.

Sedimentological evidence suggests that physical processes were indeed important in the accumulation of crinoid ossicles: cross-stratification and fining-up sequences indicate sediment transport and rapid physical events. However, since depositional textures in the present example are largely mud-supported and sorting is rather poor in most microfacies types, physical reworking was probably episodic rather than continous (cf. Cain, 1968). Clean-washed and well-sorted crinoidal limestones that, according to Cain (1968) require prolonged reworking, are rare.

Taking Hagdorn's (1978) paleoecological and the present sedimentological evidence together, massive crinoidal limestones can be inferred to have been largely produced in-situ and molded into a complex system of shallow-water bars, banks and blankets by physical processes.

Since direct modern counterparts of shallow-water crinoid sand bodies do not seem to be known, a potential analogue may be banks of muddy Halimeda-sands that occur in nearshore shallow-water environments of South-Florida (see part I). Halimeda-banks share with crinoidal banks (1) their environmental context, (2) their generally muddy depositional texture, (3) their largely in-situ accumulation of bioclastic grains and (4) the episodic impact of physical events (storms).

Bedding planes. Two main types of bedding planes can be distinguished in the massive crinoidal limestones:

a) bedding planes characterized by biological colonisation such as (1) crinoid holdfasts attached to shells, or (2) crusts, small bioherms or biostromes of the fixosessile bivalve Placunopsis ostracina (Hagdorn & Mundlos, 1982; see Fig. 33D). Because they reflect a biological response to environmental factors (erosional surface, non-sedimentation etc.), these bedding planes are clearly of primary sedimentary origin.

b) bedding planes characterized by various types of pressure solution phenomena. These commonly display crinoidal ossicles sitting on stylolite pedestals (Fig. 33E). Since crinoid ossicles appear to be more resistant against pressure solution (Logan & Semeniuk, 1976) they could survive at the top of a stylolite "front" and even serve as a template for the columnar shape of the respective stylolites. Calcitic mollusc shells (e.g. Lima) also seem to have acted as a template. Although the shell of Lima is commonly dissolved, its rib pattern is often found "telescoped" into crinoid ossicles on bedding planes (Fig. 33F). Such taphonomic evidence suggests that this type of bedding plane is at least partly of secondary, pressure solution origin ("stylo-bedding planes").

2.7. Skeletal channel fills (Fig. 34)

Description. Lensoidal or channel-like units of skeletal and intraclastic packstone, about 0.1-1 m thick and 1-3o m wide are often intercalated into alternating limestones and marlstones (Fig. 34). These units commonly show a crude fining-up sequence of skeletal packstone including several types of mostly angular limestone clasts, some of which are clearly derived from the underlying sediment while others seem to be imported. However, cross-bedded grainstone channel fills have also been observed. Most channels show on/offshore orientations (about E-W), and pebble imbrication indicates dominant offshore (W,NW) sediment transport. Several broad channels may be amalgamated to form laterally more continuous blankets of skeletal packstone.

Discussion. Stratigraphic relationships exclude a tidal flat environment and rather indicate a subtidal origin for these channels. Similar subtidal channels have been described from ancient rocks by Brenner & Davies (193), Kazmierczak & Goldring (1978) and Seilacher (1982) and were interpreted as storm surge channels or rip channels respectively. Comparable rip-scoured channels were reported by Cook (1970) from the offshore zone off California.

The multiprovenance and angularity of intraclasts and fining-up fills provides some clues regarding the dynamics of these channels. They indicate episodic rather than continuous activation and fill. Offshore directed paleocurrents document offshore flow and sediment transport consistent with a storm surge/rip/gradient current-type mechanism.

Fig. 34. Example of an on/offshore oriented channel-fill, intercalated in upper part of a shallowing-upward sequence, below skeletal shoal. Lenght of tape 1.8 m (Ilsfeld quarry).

2.8. Nodular wackestone (Fig. 35)

Description. Skeletal wackestones that display various types and degrees of nodularity are vertically and laterally associated with massive skeletal packstones and occur paleogeographically both seaward and landward of skeletal and oolitic pack- to grainstone facies belts.

These units are dark-gray coloured, extremely unsorted and strongly bioturbated, somewhat clayey biopelmicrites with various proportions of mm-thin marlstone mottles, lenses and partings. A continuous spectrum of intergradational rock fabrics can be recognized:

a) one end member is interbedded lime mudstone and graded skeletal packstone, affected by bioturbation only to a degree that primary stratification can still be well recognized and nodularity is low (Fig. 35A);

b) an intermediate stage is represented by bioturbated rocks with stronger nodularity and only few remnants of primary stratification, but with abundant pods and mottles of marlstone (Fig. 35B).

c) the other end member includes strongly nodular, totally bioturbated rocks without indication of primary stratification but with prominent marlstone lenses and partings between individual nodules (Fig. 35C).

As seen both on bedding planes (Fig. 35A) and on etched or polished rock surfaces (Fig. 35B´), bioturbation is mainly caused by large pellet-aligned spreiten-structures of Teichichnus, although other burrows such as Rhizocorallium and Thalassinoides are also present.

Discussion. Nodular limestones have been recorded by numerous authors from many different carbonate environments (e.g. Bogacz et al., 1968, Fürsich, 1973; Jenkyns, 1974; Wilson, 1975; Baud, 1976; Logan & Semeniuk, 1976; Kennedy & Garrison, 1977; Wanless, 1979b; Jones et al., 1979; Mullins et al., 1980). Many of these authors assume a combination of bioturbation and early diagenesis (submarine cementation) to form limestone nodules. In the Muschelkalk nodular wackestones, however, a combination of burrowing with late diagenesis (pressure solution) appears to be more important. Evidence for pressure solution as given by Wanless (1979b) includes the presence of microstylolites,

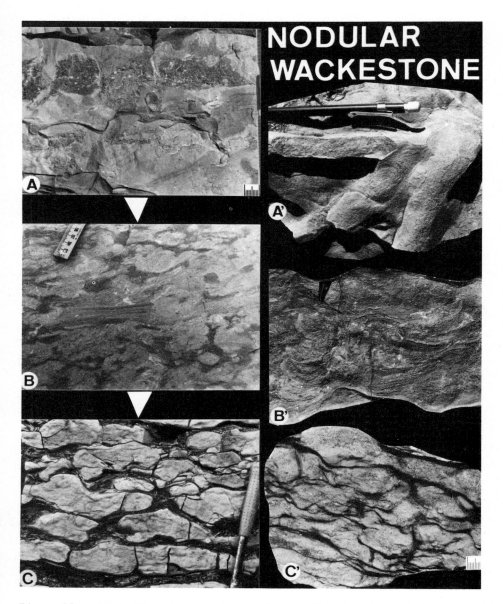

Fig. 35. Gradational spectrum of rock fabrics in nodular wackestone.
A) Interbedded thin skeletal packstone and lime mudstone, moderately
bioturbated, with well-developed individual burrows, e.g. Teichichnus
(A´). B) Mottled and nodular wackestone with only few remnants of
primary stratification but strong bioturbation with "chaos" of inter-
penetrating Teichichnus-spreiten (B´). C) Strongly nodular wackestone
with individual nodules surrounded by residual marlstone, clay seams
and microstylolites (see polished section C´). Remnants of primary
stratification and of individual burrows are practically absent.

microstylolite seams, clay seams (Fig. 35C´) and gradational top and bottom boundaries of nodules ("stylonodular rock fabric", in the terminology of Logan & Semeniuk, 1976).

The intergradational spectrum of rock types described above shows that the degree and type of bioturbation is the first factor for the development of nodularity. Pressure solution ("non-sutured seam solution" after Wanless, 1979b) is the second factor. Clay seams around nodules represent the insoluble residue along the pressure solution interfaces. Thus we deal with bioturbational inhomogenities enhanced by diagenetic processes.

2.9. Nodular lime mudstone (Fig. 36)

Description. Nodular lime mudstone is widespread in more basinal parts of the Upper Muschelkalk and is gradational to distinctly bedded limestone/marlstone alternations. It occurs in a large variety of types and ranges between relatively massive nodular micrites with thin clay seams around nodules (Fig. 36A) to isolated micritic nodules that "float" in a marlstone matrix (Fig. 36B). Nodules display a large variety of shapes (lensoid, irregular spherical, rarely somewhat angular, see Figs. 36A-E) and range in size between mm´s and dm´s. In some instances remnants of sedimentary layering and of bioturbation can be observed, but burrowing structures are generally not as easily identifyable as in the nodular wackestone facies (above). As Wanless (1979b) described, it is sometimes possible to trace sedimentary lamination from nodule to nodule through intervening zones of marlstone (Fig. 36A). The boundary between individual nodules is generally gradational and marked by anastomosing swarms of fine clay seams, microstylolites (Fig. 36C-E) and in some instances by sutured stylolites. Sharp contacts and angular limestone nodules within the surrounding marlstone matrix are rare.

Discussion. Nodular lime mudstones share so many characteristics with nodular limestones described by Wanless (1979b) that his interpretation of nodular bedding as a product of pressure solution along microstylolite seams in originally continuously layered slightly shaly limestones can be applied. This interpretation is based mainly on four characteristics:

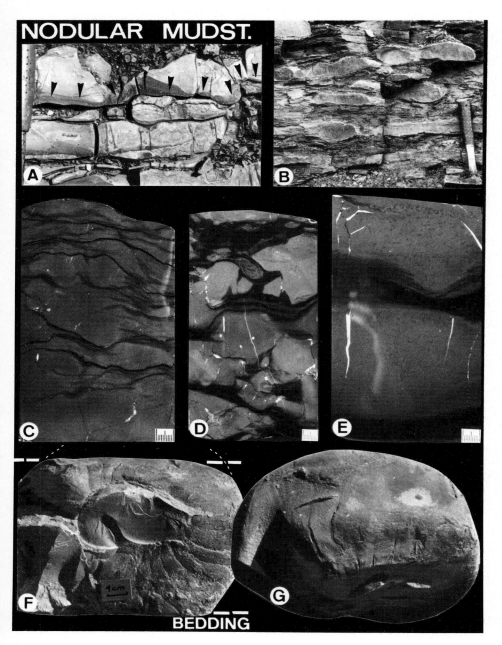

Fig. 36. Some characteristics of nodular lime mudstone. A) Slightly nodular mudstone, gradational to distinctly bedded limestone/marlstone alterations. Note that primary layering can be traced from nodule to nodule (arrows). B) Isolated nodules floating in a marlstone matrix (A,B: Dettelbach quarry). C) - E) Variability in size and shape of nodules; note that nodule boundaries are generally gradational and show swarms of clay seams and microstylolites. F) - G) Vertically embedded steinkern of _Germanonautilus_, strongly deformed by pressure solution.

a) the nodular lime mudstone facies shows continuous transitions to distinctly bedded limestone/marlstone alternations;

b) "gradational" nodule boundaries are zones of microstylolite seams;

c) tracing of sedimentary layering across several nodules provides evidence for pressure solution deformation of originally more evenly bedded layers into nodular fabrics by microstylolite formation (Fig. 36A);

d) steinkerns of ceratites or Germanonautilus (Fig. 36F,G) can be viewed as pressure solution "test bodies" of known initial size and shape. These steinkerns commonly show strong deformation by pressure solution and indicate that up to a third of their original thickness has been dissolved away.

2.1o. Graded skeletal sheets (Fig. 37)

Description. These are generally 1-30 cm thick calciruditic limestone sheets characterized by a fining-up sequence of bioclasts and intra-clasts that rests upon a sharp erosional base. The basal erosion surface commonly cuts across previously bioturbated sediment and may display gutter and trace fossil casts and tool marks, the latter being mostly bi- or multidirectional (cf. Gray & Benton, 1982). Some layers are "hummocky" (Fig. 37A), while others are more sheet-like. Hummocks are of lower amplitude (2-1o cm) and of smaller wave length (1-2 m) than the "classical" ones (e.g. Dott & Bourgeois, 1982; Hunter & Clifton, 1983).

Bedform sequences are variable. Fig. 37B shows a basal lag of imbri-cated intraclasts overlain by low-angle (hummocky) cross-stratifi-cation (HCS). An "ideal" sequence is represented in Fig. 37C: sharp, erosional base, followed by a fining-up shelly layer, overlain by parallel, then low-angle lamination, topped by wave-ripples. Wave-ripples are particularly common at bed tops (Fig. 37E). They are often slightly asymmetrical and have ripple indices from 7 to 16 thus re-presenting largely low-steepness ripples (e.g. Allen, 1981). Clear-cut current ripples have not been observed. Many beds are composite and in-clude the amalgamation of several depositional events (or episodes within one event, Fig. 37D). Such composite sheets may be internally homogeneous at one locality, but when traced laterally for just a few meters, the same bed may be composed of a number of individual, amalgamated fining-up layers, separated by discrete bedding planes.

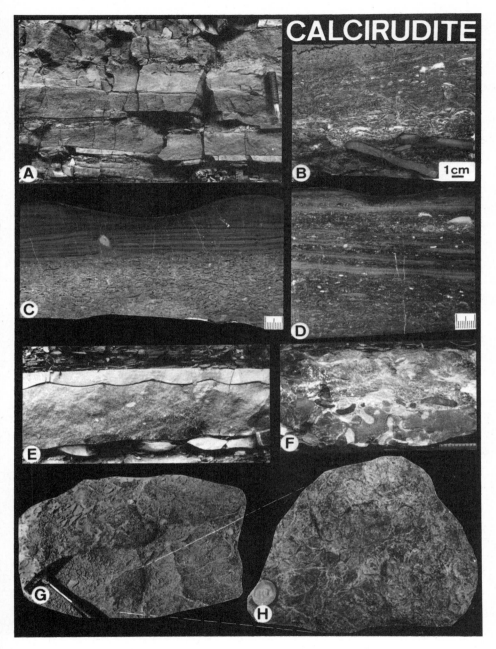

Fig. 37. Stratification types in graded calcirudite sheets. A)Proximal
tempestites with HCS. B) Slab of proximal tempestite. Note imbricated
intraclasts at base, low-angle stratified (HCS) graded sequence and
bioturbated top. C) "Ideal" tempestite: sharp base, graded coquina,
passing into parallel then low-angle lamination, wave ripples at top.
D) Composite bed with several depositional increments. E) Graded skeletal packstone with wave-rippled top. F) Burrowed high-relief firmground with black pebbles overlain by a calcirudite bed. G) - H) Large
reworked limestone slabs, encrusted by Placunopsis, at the base of a
thick calciruite bed.

The bases of many such composite beds display high-relief firmgrounds and large reworked pebbles (Fig. 37F-H). Borings and encrustation of these pebbles from several sides attest to repeated reworking and colonisation (Fig. 37G,H). Bed tops are commonly bioturbated by Teichichnus, Rhizocorallium irregulare, and Planolites/Palaeophycus.

Discussion. These beds have previously been described and interpreted as proximal storm sheets or tempestites by Aigner (1979, 1982, 1984) and Mehl (1982). Very similar sequences have been recorded from numerous ancient clastics and carbonate shelf deposits (summaries in Goldring & Bridges, 1973; Ager, 1974; Brenchley et al., 1979; Kreisa, 1981; Dott & Bourgeois, 1982; Einsele & Seilacher, 1982). Modern analogues include storm layers in Recent continental shelves such as from the Gulf of Mexico (Hayes, 1967; Morton, 1981), Bering shelf (Nelson, 1982) and the southern North Sea (part I, Aigner & Reineck, 1982).

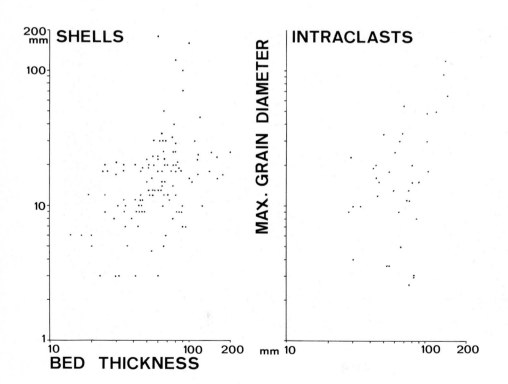

Fig. 38. Plot of maximum diameters of shells and intraclasts against thickness of calcirudite beds.

The particular association of sedimentary structures in these beds allows the _dynamics_ of their formation to be reconstructed. Firstly, the vertical succession of structures (e.g. in the "ideal" sequence, Fig. 37C) indicates deposition during waning flow. Secondly, the presence of both wave-formed structures (bi/multidirectional tool marks, wave ripples, infiltration and grain-sheltered fabrics) and of features caused by currents (lateral sediment transport, primary current lineation) suggests a complex interaction of oscillatory and unidirectional flows. Such _combined flows_ can be expected during storms when storm waves touch the sea floor and interact with uni-directional currents (e.g. gradient currents).

As is found in limestone turbidites (e.g. Sadler, 1982), tempestite thickness varies proportionally with maximum grain size of the bed (Fig. 38); it can therefore be inferred, that the thickness of tempestites is controlled by the fall velocity of the largest grain carried to a given point and must be related to the shear stress exerted by the flow (cf. Sadler, 1982).

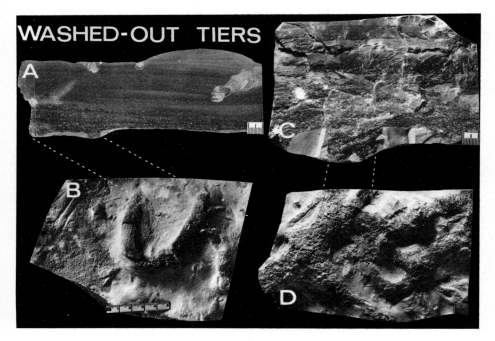

Fig. 39. Preservation of trace fossil tiers at the sole of tempestites. A) Cross-section of _Rhizocorallium_ washed out and cast by a tempestitic calcarenite bed. B) Cast _Rhizocorallium_, viewed from bed base. C) Cross-section of _Thalassinoides_ burrows washed out and cast by graded calcirudite bed. D) Same specimen as in C), view of bed base with _Thalassinoides_ network.

Bedding planes. The bases of calcirudite sheets (proximal tempestites) are always erosional, with scours, tool marks and cast trace fossils (Fig. 39B). Washed out and cast trace fossils can be used as a measuring stick for the minimal depth of storm erosion. This is possible because trace fossils are commonly tiered, i.e. they show a characteristic vertical zonation within the sediment (Wetzel, 1979; Ausich & Bottjer, 1982). In the Muschelkalk, Rhizocorallium irregulare belongs to a shallow tier, while Thalassinoides occurs deeper in the sediment. If Rhizocorallium is washed out and cast by an overlying tempestite, seafloor erosion was only in the order of a few cm, in contrast to cast Thalassionides, that indicates deeper erosion of more than 10 cm (Fig. 40). This way, the minimal amount of sediment lost by erosion and represented by a bedding plane can be quantified. When applied to an entire sequence, this method allows to determine the degree of stratigraphic incompleteness on a bed-by-bed base (Fig. 41).

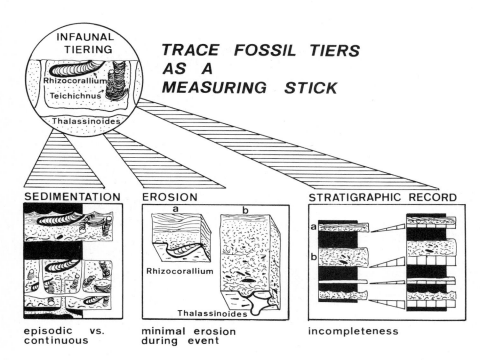

Fig. 4o. Infaunal tiering in trace fossil associations provides a measuring stick for the dynamics of stratigraphic accumulation: (1) undisturbed versus interpenetrated tiers indicate episodic versus continuous sedimentation (left). (2) Washed out and cast trace fossils below event deposits indicate minimal erosion during the event (middle). (3) Thus stratigraphic incompleteness can be determined quantitatively (right).

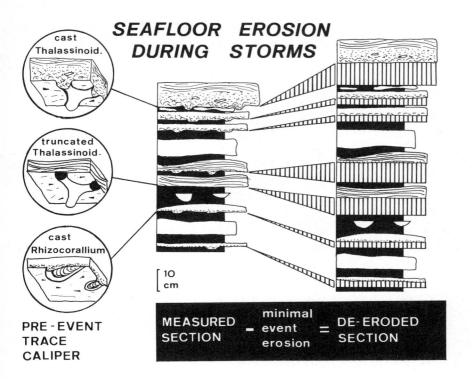

Fig. 41. Representative section in limestone/marlstone alternation and some common styles of trace fossil preservation at the base of limestone beds. Because Thalassinoides belongs to a deep infaunal tier, cast or truncated Thalassinoides burrows indicate deep event erosion (> 10 cm). Rhizocorallium, in contrast, is typical for a shallow tier, hence cast Rhizocorallium burrows indicate only minor erosion in the order of 1-3 cm. Subtraction of minimal erosion values (deduced from trace fossil preservation) from the measured section gives "de-eroded section" and quantitative values for stratigraphic shortening. Stratigraphic incompleteness of this sequence is about 1/3.

The top surfaces often show evidence for biological colonisation following the event ("post-event colonisation"). Such surfaces are of three main types (Fig. 42): a) firmgrounds burrowed by Glossifungites (and Thalassinoides); b) shellgrounds colonized by brachiopods (e.g. Spiriferina fraglis) and byssally attached bivalves (Pleuronectites, Lima); c) hardgrounds bored by Calciroda, Talpina and encrusted by Placunopsis and terquemiid oysters. These types of surfaces clearly represent primary biological colonisation "events" and may thus be called "event bedding planes". Many other bedding planes, particularly in slightly argillaceous lime mudstones, are strongly obliterated by stylolites and microstylolites. In these cases, fossils are highly distorted and often hardly recognizable (Fig. 48); such surfaces may be called "stylo-bedding planes".

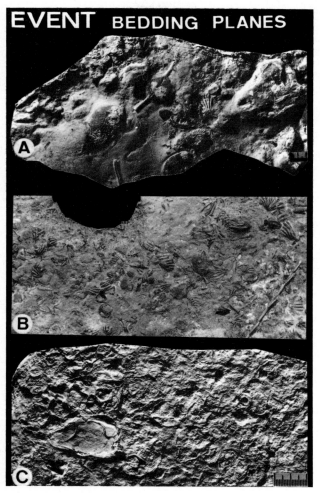

Fig. 42. Post-event colonisation of tempestite tops indicating "event bedding planes". A) Firmground burrowed by Glossifungites and colonized by pectinids and oysters. B) Shellground colonized by brachiopods (here Spiriferina fraglis). C) Hardground encrusted by Placunopsis ostracina and bored by Calciroda.

2.11. Thin-bedded limestone/marlstone alternations (Figs. 43-48)

Description. This lithology is most widespread in the more central
parts of the Upper Muschelkalk Basin ("Tonplatten-Fazies"): 1-10 cm
thin slightly argillaceous calcarenite, calcisiltite and calcilutite
sheets alternate with cm-thin marlstone layers (argillaceous calcilutites, see Fig. 43A). Bed geometries include lenses and wedges and
small "hummocks" (wavelength a few dm, amplitude 1-3 cm) but are
dominated by slightly irregularly bedded limestone sheets.

Bed bases of calcarenites/calcisiltites are always erosional and often
display several types of tool marks (bounce and bipolar prod marks,
Fig. 44) as well as casts of washed-out trace fossils (Fig. 39A). Many
beds have a cm-thick micritic layer attached to their erosive base
(e.g. Fig. 43F); these "underbeds" are diagenetic aureoles below event
surfaces (cf. Aigner, 1982a; Eder, 1982). As with calcirudite sheets
(section 2.10), bedform sequences are also variable in the calc-
arenite/calcisiltites. Fig. 43B shows an "ideal" succession: sharp,
erosional base, overlain by a thin skeletal lag, followed by parallel
lamination that develops upward into low-angle lamination, topped by
wave ripples. There are many modifications from this "ideal" motive
such as undulous and climbing wave ripple laminations (Fig. 43C),
distinct fining-upward sequences (Fig. 43D), or graded rhythmites
(Fig. 43E; see Reineck & Singh, 1972). Most common, however, are
faintly laminated units (Fig. 43F). Composite beds, that include
several depositional increments may also occur (Fig. 43G). In rare
instances, calcisiltites show evidence for synsedimentay deformation
(convolutions, ball-and-pillow structures, slumping; see Fig. 45).

Under the SEM, badly preserved tests of the nanno-organism
Schizosphaerella have been recorded in some samples (Fig. 46).

In outcrops, most of the calcilutite beds appear structureless except
for bioturbation at bed tops (Fig. 47A,B). Other layers are coarser
near the base and display thin lags of shell debris (Fig. 47C,D) or
zones of laminated calcisiltite (Fig. 47E). Many other beds, however,
are internally homogenous lime mudstones with undulating gradational
boundaries to the under- and overlying marlstones and without apparent
sedimentary structures, except occasional burrows (Fig. 47F,G). This
facies is gradational to nodular lime mudstones.

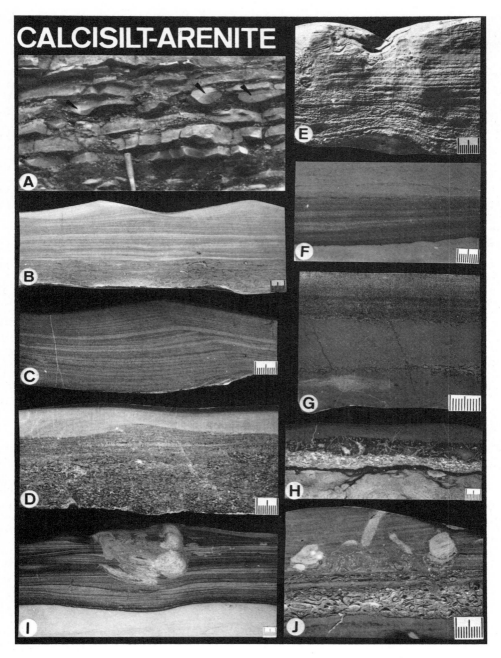

Fig. 43. Stratification in thin-bedded calcarenites/calcisiltites. A) Thin-bedded limestone/marlstone alternations with gutter casts (arrows). B) "Ideal" tempestite: sharp base, skeletal lag, parallel to low-angle lamination, wave-rippled top. C) Climbing wave-ripple lamination . D) Graded calcarenite. E) "Graded rhythmite". F) Laminated calcisiltite with "underbed". G) Composite bed with three depositional increments. H) Calcarenite layer with blackened and bored hardground at top. I)-J) Post-event burrows: I) Teichichnus, J) Rhizocorallium.

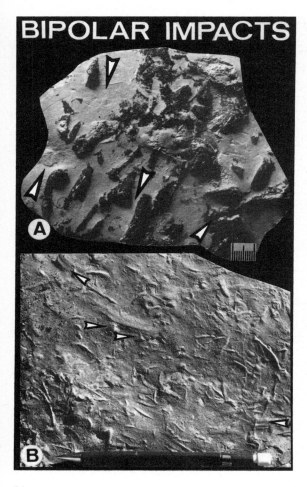

Fig. 44. The soles of many tempestites are characterized by bounce marks and prod marks the latter typically with bipolar orientations (arrows). Bipolar prod marks emphasize the oscillatory flow component during storm erosion and form an important criterion to distinguish tempestites from other types of event deposits (e.g. turbidites with unidirectional impacts).

<u>Discussion</u>. In <u>calcarenites</u> and <u>calcisiltites</u>, sedimentary structures and their succession are similar to those in the calcirudite sheets (section 2.10); a similar storm-induced depositional mechanism can therefore be inferred. Sharp bases and internal bedform sequences suggest rapid erosion and deposition during one event. Oscillatory flow components are indicated by bi- or multidirectional prod marks and by wave ripples. Lateral and vertical transitions into the thicker-bedded calcirudite sheets demonstrate that these thinner-bedded calcarenite/calcisiltite sheets are the "distal", deeper water equivalents to "proximal" tempestites (cf. Aigner, 1982a).

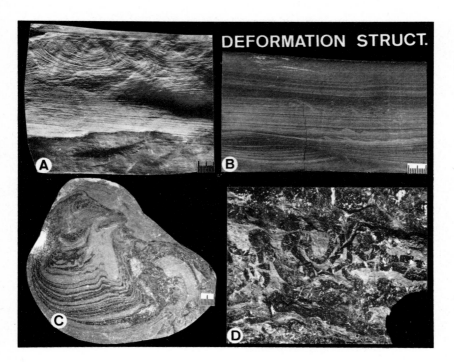

Fig. 45. Soft-sediment deformation structures are rare in the Upper Muschelkalk and have only been observed in calcisiltites and calcilutites. A) Convolute lamination in upper part of tempestite. B) Minor convolutions in fill of gutter cast. C) Isolated "ball and pillow". D) Slumping of thin calcilutite bed (diameter of camera cap = 5 cm)

In the calcilutites, the fact that bioturbation always starts at bed tops documents also instantaneous deposition of the entire bed during one event, followed by episodic post-event colonisation. Event-deposition is also indicated by sharp bed bases and basal lags. The facies context and transitions to calcarenitic/calcisiltitic tempestites suggest that these thin and fine grained layers are the most distal "fine tails" of storm flows in offshore deeper water. Thin calcilutites that occur together with "proximal tempestites" may represent spills from "proximal" storm surge channels, analogous to overbank turbidites in the deep sea. Very similar storm-generated silt-stones and silty mudstones have been described by Brett (1983). "Mud-tempestites" from the Helgoland Bight are possible modern counterparts (see part I; Aigner & Reineck, 1982).

Whether the structureless layers with gradational boundaries are also event-deposits or whether they represent "diagenetic bedding" in the sense of Ricken (1985) remains an open question.

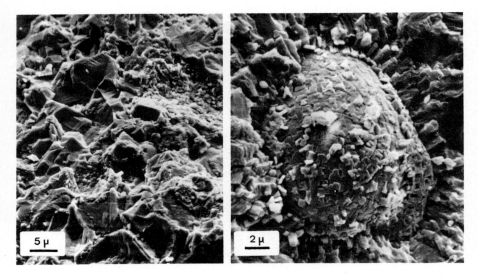

Fig. 46. In few samples of calcisiltites, badly preserved tests of Schizosphaerella have been recorded under the SEM. Schizosphaerella is a presumably planctonic nanno-organism of unknown systematic position and occurs widely throughout the Jurassic (Kälin, 1980) but does not seem to be known from older rocks as yet. Its role as potential contributor to fine-grained carbonate should be further evaluated, but is hampered by recrystallisation of micrites into microsparite.

Bedding planes. Bed tops are commonly colonized by infaunal organisms, either by borers (hardgrounds, Fig. 43 H) or by burrowers (Teichichnus, Fig. 43I; Rhizocorallium, Fig. 43J). Some bed tops show shell pavements (Aigner, 1977), some firmgrounds, burrowed by Balanoglossites and Glossifungites (Fig. 47C-D).

Many bedding planes are strongly affected by pressure solution (stylolites and microstylolites). Fossils and their degree of distortion can be used as "test bodies" to estimate the amount of pressure solution on these stylo-bedding planes. Fig. 48 shows various stages of distortion in (1) ceratite steinkerns (Fig. 48A,A´), in (2) Planolites/Palaeophycus burrows (Fig. 48B,B´,B´´), and in (3) ophiurids on bedding planes (Fig. 48C,C´). Due to pressure solution, sediment loss on such "normal" bedding planes is often in the order of several millimeters or even more.

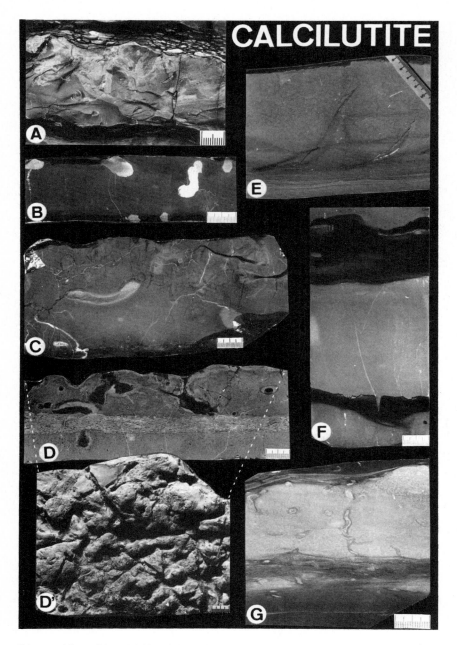

Fig. 47. Stratification in thin-bedded calcilutites. A) Field appearance of calcilutite bed with bioturbation from top. B) Unspecific burrows (?Balanoglossites, cf. Kazmierczak & Pszczolkowski, 1969) at bed top. C) Thick calcilutite bed with veneer of skeletal debris at base and intense bioturbation from top. D) Calcilutite bed with skeletal lag and well-developed firmground with Glossifungites-burrows at top, D´) top view of firmground. E) Homogenous calcilutite bed with thin laminated calcisiltite at base. F)-G) Homogenous calcilutites in cores: F) with relatively sharp boundaries, G) with gradational boundaries and some burrows; note strong compaction of burrows in marl.

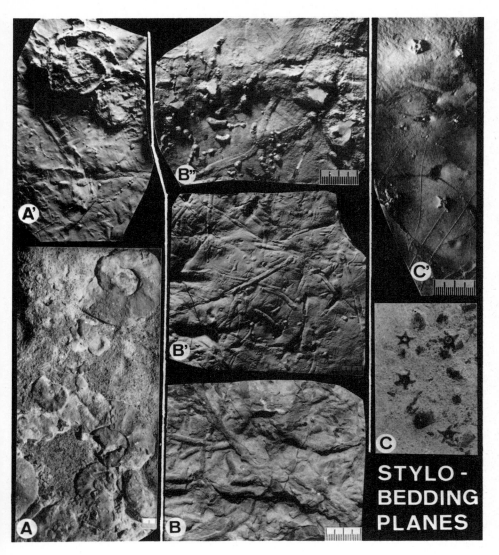

Fig. 48. Fossils as test-bodies to deduce degree of pressure solution on "stylo-bedding planes". A) Steinkerns of ceratites, relatively un-affected by pressure solution; A´) "Ruins" of ceratite steinkerns caused by strong stylolithization. B) Undistorted Planolites/Palaeophycus burrows, B´) to B´´) increasing degree of distortion by pressure solution. C) Unaffected ophiurids (<u>Aspidura</u> sp.); C´) Stylolithized remnants of ophiurids on pressure solution affected bedding plane (both specimens courtesy of H. Hagdorn).

2.12. Conclusions: storm-dominated stratification

This chapter aimed to (1) analyze the most important stratification and facies types of the Upper Muschelkalk and (2) answer a set of questions concerning the "stratinomy" of these deposits (see section 2.1.):

1. Partly based on actualistic analogues, most stratification types in the Upper Muschelkalk show evidence for episodic storm events and can be used as <u>environmental indicators</u>. Supratidal storm layers indicate "storm flats" (according to Wanless, 1976), and thin fining-up layers in nodular back-bank deposits record episodic storm reworking. In analogy to modern shallow-water carbonate sand bodies, massive oolitic, shelly and crinoidal pack- to grainstones are likely to have been molded and activated in response to storm flows. Channel-fills seem to record episodic erosion by storms. Vertical and lateral transitions from calciruditic into calcarenitic/calcisiltitic and calcilutitic storm sheets (tempestites) demonstrate that these re-present a continuous spectrum of genetically related rock types ranging from "proximal" (generally shallow, nearshore) to "distal" (generally deeper, offshore) end members (Fig. 49). Proximal to distal changes are expressed in decreasing bed thicknesses, degree of bed amalgamation, grain size, bioclast and intraclast content, and in changing sedimentary structures, paleocurrents and faunas. Such <u>proximality trends</u> are similar to trends in modern storm layers (see part I; Aigner & Reineck, 1982; Nelson, 1982) and reflect the decresing effects of storms towards deeper, offshore water. They can thus be used for facies and paleobathymetric zonations.

2. Aspects of the <u>depositional dynamics</u> of tempestites can be reconstructed based on stratification characteristics (Fig. 50):

a) The <u>minimal amount of storm erosion</u> at the seafloor can be esti-mated by (1) the types of trace fossils cast at tempestite bottoms, and (2) the burrowing depth of reworked versus non-reworked infaunal shells. The first method is possible because trace fossils are commonly tiered, i.e. they show a characteristic vertical zonation of burrows within the sediment. The deep-burrowing <u>Thalassinoides</u> are commonly cast by proximal tempestites and reflect deeper storm erosion in the order of perhaps a few dm. In contrast, the shallow burrowing <u>Rhizocorallium</u> is often cast by distal tempestites, reflecting minor

TEMPESTITE PROXIMALITY

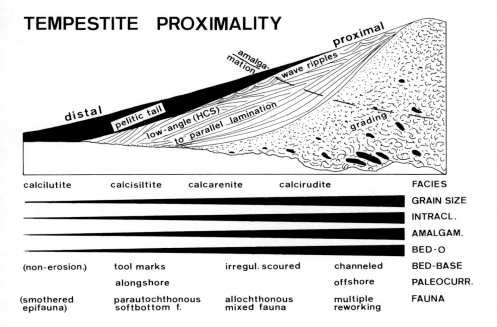

				FACIES
calcilutite	calcisiltite	calcarenite	calcirudite	
				GRAIN SIZE
				INTRACL.
				AMALGAM.
				BED-Ø
(non-erosion.)	tool marks	irregul. scoured	channeled	BED-BASE
	alongshore		offshore	PALEOCURR.
(smothered epifauna)	parautochthonous softbottom f.	allochthonous mixed fauna	multiple reworking	FAUNA

Fig. 49. Generalized trends in vertical and lateral variation of "ideal" tempestite: proximality as a guide for environmental interpretations.

erosion in the order of a few cm. Since the deeply burrowing bivalve _Pleuromya_ is normally not reworked into tempestites, the average depth of storm erosion on the seafloor was probably in the order of cm's. Such values are compatible with erosion rates found during modern storms (Kolp, 1958; Smith & Hopkins, 1972; Kumar & Sanders, 1976).

b) The particular association of sedimentary structures allows speculations on the _kind of storm flows_. Wave ripples and bi-directional tool marks represent oscillatory flow components, while other features (e.g. lateral sediment transport, imbrication) document superimposed unidirectional flows. Combined flows can thus be inferred, although the complete mechanism is not yet fully understood. Possibly wave erosion was most important during initial storm phases (bipolar sole marks), followed by a dominance of unidirectional flows (lateral sediment influx), before oscillatory flows became again dominant during storm waning (wave-rippled tops).

c) The _direction of storm flows_ is indicated by the orientation of tool marks at bed bases, imbrication within the bed, and wave ripples at bed tops. Prod marks at the bases of distal tempestites (and gutter casts) show bipolar longshore orientations, and wave ripples indicate longshore winds accordingly. Imbrication of intraclasts and shells in proximal tempestites record largely offshore directed flows.

d) The grain size in tempestites allows some estimates on the order of magnitude of <u>minimal storm flow velocities</u>. In the Muschelkalk, "distal" calcarenites/calcisiltites seem to have been deposited by low-velocity flows in the order of 1-20 cm/s. These inferences are based on flume experiments of Miller & Komar (1977) and Wanless et al. (1981) with foraminiferal tests and with pellets. More "proximal" shelly calcirudites were probably deposited by flows in the order of 20-60 cm/s, based on flume experiments on shells by several workers (Trusheim, 1931; Johnson, 1957; Futterer, 1978). Very coarse calci-rudites may indicate maximal flow velocities of 60-400 cm/s using maximal pebble size and the critical erosion velocity-entrainment curves of Sundborg (1967). Similar approaches are commonly used to de-duce paleovelocities in fluvial sedimentology and should principally be applicable to storm layers. However, complications arise from the poorly understood nature of combined flows. In any case, the flow velocities inferred for the Muschelkalk tempestites are well in agree-ment with flow velocities measured during present-day storms (e.g. - 200 cm/s, Forristall et al. 1977; - 151 cm/s, Gienapp, 1973; -80 cm/s, Sternberg & Larsen, 1976; - 50 cm/s, Cacchione & Drake, 1982; 20-60 cm/s, Swift et al., 1983).

e) <u>Wave and sea conditions</u> can theoretically be reconstructed based on grain size and the morphology of wave ripples on tempestite tops using approaches of Bourgeois (1980), Hunter & Clifton (1982), Sundquist (1982) and P.A. Allen (1984). In the present case, such calculations are difficult, however, because (1) the hydrodynamic behaviour of carbonate grains is not yet sufficiently understood, (2) most of the present ripples have a vertical form index > 7.5 that should not be used for wave reconstructions (P.A. Allen, 1984), and (3) superimposed unidirectional currents are likely to also seriously affect such calculations (P.A. Allen, 1981).

3. Storms are important for the <u>ecological dynamics</u> on the seafloor and are recorded in some characteristic paleoecological patterns. Analogous to recent shallow water environments (e.g. Schäfer, 1970; Rees et al., 1977; Shackley & Collins, 1984), storms play a part in both the destruction and the formation of benthic associations: storm

scour washes out and accumulates soft-bottom faunas to form shell
layers (e.g. Aigner, 1977), yet at the same time creating substrates
for new firm-, shell- and hardground associations as a biological
response ("taphonomic feedback", Kidwell & Jablonski, 1983). In the
Upper Muschelkalk, proximal tempestites typically include mixed,
partly transported faunas, while distal tempestites are dominated by
in-situ reworked parautochthonous assemblages, similar to North Sea
storm shell beds (see part I, Fig. 24).

Post-event bioturbation commonly reworked the upper part of, sometimes
even the complete storm-generated layers. Since bioturbation in many
distal, basinal sequences is restricted exclusively to tempestite
tops, it may be inferred that benthic colonisation of the seafloor was
at times inhibited during background conditions (oxygen, salinity
control ?), but only possible episodically after storm events.

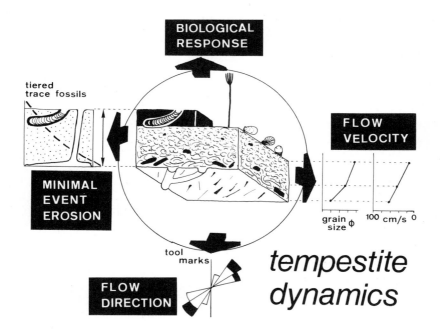

Fig. 50. Summary of some dynamic factors that can be deduced from
tempestites: (1) The preservation of trace fossil tiers at the base of
tempestites allows to estimate the amount of minimal erosion during
the event. (2) Tool marks at tempestite soles indicate the direction
of storm flows. (3) The grain size of tempestites gives some indication on storm flow velocities. (4) The type of post-event faunas
on tempestite tops sheds light on paleoecological factors such as
substrate consistency and community structure.

4. Primary bedding may be modified by <u>pressure solution</u> both in form of "sutured seam solution" (Wanless, 1979b) most common in grain-supported shallow water facies, and in form of "non-sutured seam solution" (Wanless, 1979b) in more argillaceous, basinal rock types. For instance, pressure solution - together with bioturbation - has transformed primarily well-stratified (storm-stratified) facies into a variety of "nodular" rock fabrics.

5. <u>Bedding planes</u> are of two basic types: a) event bedding planes that record erosional (e.g. scoured bed bases with tool marks) or depositional surfaces (e.g. tempestite tops available for biological colonisation), and b) stylo-bedding planes that are modified or entirely caused by pressure solution. Many "normal" bedding planes represent significant gaps in the stratigraphic record caused by syndepositional erosion and non-deposition as well as by postdepositional pressure solution.

3. FACIES SEQUENCES

3.1. Vertical sequences: coarsening-upward cycles

Detailed bed-by-bed analysis of the Upper Muschelkalk reveals a marked and pervasive sedimentary cyclicity (Fig. 51). Individual cycles (mostly between 1 and 7 m thick) show the following trends in upward direction: 1. coarsening of grain size, 2. thickening of beds, 3. increase in physical sedimentary structures, 4. lighter colours, and 5. faunistic changes. From the variety of sequences following one or more of these general motives, some of the most common types are discussed here. Cycles with the opposite trends (e.g. fining- and thinning-upward) are rare in the study area, but have been recorded in the deep basinal Main region (see Fig. 76).

3.1.1. Oncolitic cycles (Fig. 52)

Description. This type of sequences is only present in very marginal parts of the Upper Muschelkalk Basin and occurs paleogeographically mainly landward of oolite grainstone cycles.

The basal parts of these sequences comprise dark, rather argillaceous, often nodular lime mudstone, marlstone and bioturbated (Teichichnus) peloidal wackestone (Fig. 52A), in some cases with abundant small black pebbles and blackened shell debris. Upwards, marlstone layers become fewer and thinner. The proportion of micritic envelopes around shells increases and some bio- and lithoclasts may show incipient algal coatings and small oncolites (Fig. 52B). The upper portion of such sequences is commonly formed by a 1/2 - 2 m thick bed of oncolite-rich skeletal packstone (Fig. 52C). These sometimes cross-bedded units may be lensoidal and channel-like or form more sheet-like sediment bodies.

Discussion. The upward change from bioturbated and unsorted fine-grained mud- and wackestone to better sorted and stratified coarser pack- and grainstone indicates an upward shallowing trend. The paleogeographic distribution of this type of cycle suggests a shallow, somewhat restricted, "lagoonal" depositional environment. This inference

SEDI-MENTARY CYCLES

Fig. 51. Many different types of asymmetrical coarsening-upward sedimentary cycles can be readily observed in weathered quarry sections. Most typical are : A) Oolite grainstone cycle (see Fig. 53). B) Skeletal bank cycle (see Fig. 54). C) Crinoidal bank cycle (see Fig. 55). D) Nodular-to-compact cycle (see Fig. 56). E) Thickening-upward cycle (see Fig. 57). F) Thinning- and fining-upwards cycles are rare in the study area.

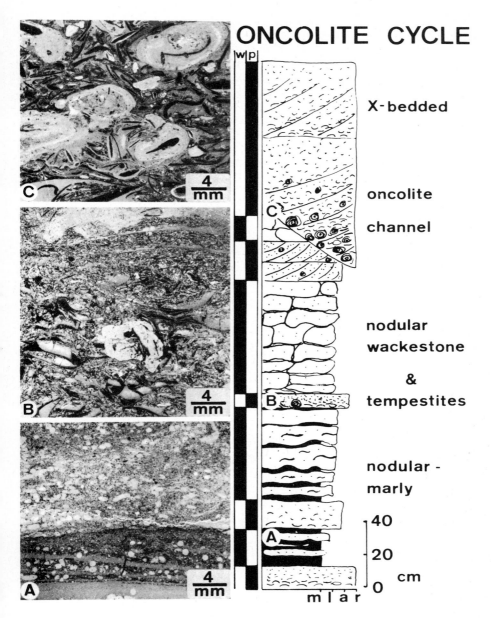

ONCOLITE CYCLE

w|p

X-bedded

oncolite

channel

nodular
wackestone

&

tempestites

nodular -
marly

40

20

cm

0

m l a r

Fig. 52. Detailed log through an oncolite cycle with upward change in microfacies from nodular pelmicrite (A), to thin graded calcarenites-calcirudites (B), to massive oncolitic calcirudites.
Explanation of symbols for all following logs: M = mudstone, W = packstone, P = packstone, G = grainstone; m = marl, l = calcilutite, a = calcarenite, r = calcirudite.(section no. 6, Tiefenbach)

is also supported by (1) abundant black pebbles, that are typical for shallow nearshore settings (Barthel, 1974; Strasser & Davaud, 1983), and (2) oncolites, that occur preferentially in ponds and channels within back-bank environments (Wilson, 1975; see also Wagner, 1913b). These cycles are therefore interpreted to represent lagoonal areas gradually shoaling into oncolitic ponds and channels.

3.1.2. Oolite grainstone cycles (Fig. 51A, 53)

Description. This type of sequences is most common in a prominent belt in marginal parts of the Muschelkalk Basin and reaches generally between 2 and 6 m in thickness. The basal part is normally dark-grey lime mudstone and pelmicrite (Fig. 53A) commonly with a pronounced nodular fabric and marly partings. Nodularity is caused largely by ubiquitous bioturbation, particularly by large pellet-filled spreiten-structures of Teichichnus. Remnants of laminations are only rarely preserved.

The basal part grades upwards into lighter-grey unsorted wackestones and packstones (intrabiomicrites) that are thicker bedded and without or with only minor marlstone partings (Fig. 53B). The degree of bio-turbation and nodularity is still high, although remnants of primary stratification such as cm-thick discontinuous fining-up sequences are more common. As lime mud content decreases upwards, the degree of sorting and the abundance of micritic envelopes around shells increases.

The upper part of these cycles is usually a 1-3 m thick unit of light grey to yellowish cross-bedded oolitic grainstone (Fig. 53C). This is usually well-sorted; most bioclasts are rounded and have micritic envelopes. Evidence of submarine and vadose diagenesis (dripstone and meniscus cement) is present. The top surface of these cycles is generally sharp and in a few instances marked with thin Fe-veneers and rare hardgrounds. In Muschelkalk lithostratigraphy, many such cycle-top units have been named (e.g. "Obere Oolithbank") and used for correlation, although the general cyclicity was not recognized.

Discussion. All patterns of progressive change within this type of sequence indicate shallowing-upward cycles. They correspond to Wilson's (1975) "asymmetric shoaling-up carbonate shelf cycles". Analogously they are interpreted as the result of a rapid rise in

OOLITE-GRAINSTONE CYCLE

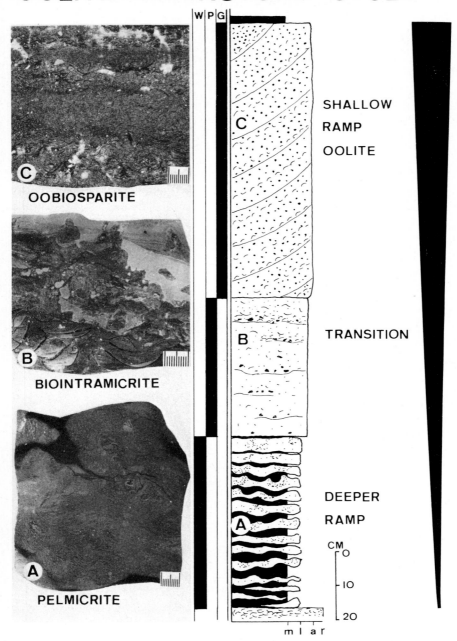

OOBIOSPARITE

BIOINTRAMICRITE

PELMICRITE

W P G

C

SHALLOW
RAMP
OOLITE

B

TRANSITION

A

DEEPER
RAMP

CM
0

10

20

m l a r

Fig. 53. Detailed log through oolite grainstone cycle (cf. Fig. 51A).
Note upward changes from deep-ramp bioturbated pelmicrites (A) through
weakly bioturbated skeletal wackestone and packstone (B) into cross-
bedded shallow ramp oolitic grainstone (C). Section no. 21 (cf. Fig.
73), Wilhelmsglück (from Aigner, 1984).

relative sea level followed by regressive seaward outbuilding of the
oolitic shoalwater complex. Cycle tops document periods of non-deposition and in some cases local subaerial exposure.

3.1.3. Skeletal bank cycles (Fig. 51B, 54)

Description. These sequences occur mainly seaward, but also landward
of the belt with oolitic grainstone cycles in the marginal limestone
facies province of the Upper Muschelkalk.

The vertical succession of sedimentary structures, microfacies and
trace fossils is variable but generally similar to oolite grainstone
cycles (except for the ooids). Commonly, sequences pass from
argillaceous, thin- to nodular bedded wackestone (Fig. 54A) to
thicker-bedded pack- to grainstone (Fig. 54B), that is overlain by a
thick (1-3m) skeletal grainstone at the top (Fig. 54C). Just below
these massive top units, skeletal channel fills as described above
(section 2.7) may be present. The uppermost massive bed is commonly
cross-bedded, well-sorted and includes a high proportion of micritic
envelopes. As in oolite grainstone cycles, many such units form marker
beds (e.g. "Mittlere Schalentrümmerbank") in Upper Muschelkalk lithostratigraphy.

Discussion. As in the oolite grainstone cycles, all patterns of upward
change indicate upward shallowing sequences. Basal argillaceous and
nodular limestones represent deeper water conditions that pass into
tempestites and thin sheets of skeletal sand. Channelized beds are
shallow-water storm surge channels and the uppermost massive skeletal
unit is interpreted to represent a very shallow shoalwater complex of
skeletal bars and banks.

3.1.4. Crinoidal bank cycles (Fig. 51C, 55)

Description. This type of sequence occurs only in the lower part of
the Upper Muschelkalk ("Trochitenkalk") and is most prominently
developed in more marginal parts of the basin.

Sequences start with basal argillaceous units; in the figured example
(Fig. 55) the basal part is in fact a thick marlstone horizon with few

SKELETAL BANK CYCLE

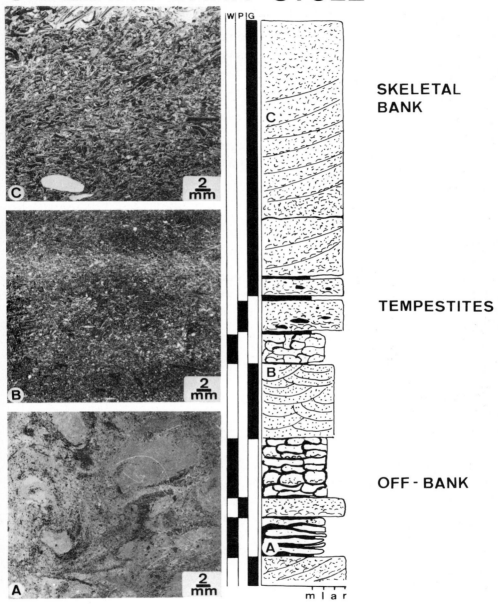

SKELETAL
BANK

TEMPESTITES

OFF - BANK

Fig. 54. Detailed log through a skeletal bank cycle (cf. Fig. 51B) with upward change from bioturbated arenitic wackestone (A), to well-sorted arenitic grainstone with few micritic envelopes (B), to well-sorted ruditic grainstone with abundant micritic envelopes. Section no. 5, Neidenfels (cf. Fig. 72).

CRINOIDAL BANK CYCLE

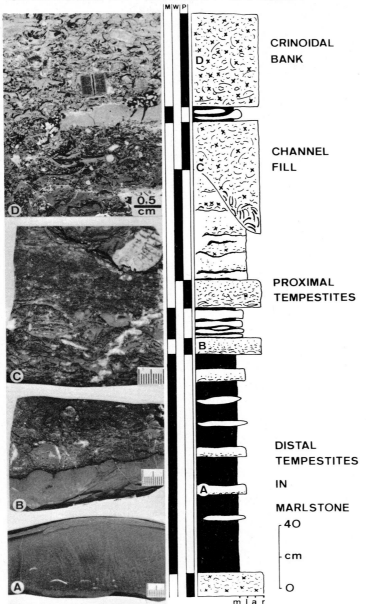

Fig. 55. Detailed log through a crinoidal bank cycle (cf. Fig. 51C) with upward change from thin graded calcisiltite (A, distal tempestite), to graded shelly sheets (B, proximal tempestite), to channeled skeletal packstone (C), to massive shelly/crinoidal limestone (D). Section no. 37, Bretten (cf. Fig 71).

thin limestone nodules. Upwards, increasingly more and thicker limestone layers are interbedded within the marlstone. These limestones
represent typical "distal" tempestites (Fig. 55A) and pass further
upwards into "proximal" ones (Fig. 55B), which may include some
crinoid ossicles, in contrast to the distal ones. Within the upper
part of these sequences, small skeletal channel fills as described
above (section 2.7.; Fig. 55C) may be present and include large
amounts of crinoid ossicles. The top unit of these cycles is represented by a 1/2 - 2 m thick mostly massive, and in some cases
oolitic crinoid-mollusc-brachiopod pack- to grainstone (Fig. 55D).

The benthic macrofauna and the ichnofauna show also marked changes
through these cycles. While their lower part is dominated by soft-
bottom assemblages and burrows of sediment feeders (_Rhizocorallium_
irregulare), the upper part includes firm-, hard- and shellground
faunas and burrows of suspension feeders (e.g. _Glossifungites_).
Complete specimens of the crinoid _Encrinus liliformis_ are preferentially found at cycle tops (e.g. Linck, 1965, in "Trochitenbank 6").

Discussion. The vertical transition from marlstones to distal and
proximal tempestites, to massive skeletal units indicates upward
shallowing. Microfacies and faunal changes support this conclusion.

The restriction of articulated and in-situ crinoids and of some other
fixosessile suspension feeders to the upper part of these cycles is
most probably the ecological response to shallowing (regression). Only
a higher energy level caused extensive winnowing that provided the
firm, shelly or hard substrate necessary for the attachment of
crinoids and other fixosessile epibenthos. Hence it was only during
general shallowing that the crinoids could spread out from their more
permanent habitats on swells (cf. Hagdorn, 1978, 1985) and colonize
larger parts of the basin during peak regression. The thus formed
crinoid-rich massive skeletal units have long been used as lithostratigraphic markers (e.g. "Trochitenbank 6").

3.1.5. "Nodular-to-compact cycles" (Fig. 51D, 56)

Description. Paleogeographically, this type of sequence is common in
an intermediate area between the nearshore shoalwater complex and more
basinal settings, but may also occur landward of the oolite grainstone
cycles. On weathered sections, "nodular-to-compact cycles" are easily

NODULAR-TO-COMPACT CYCLE

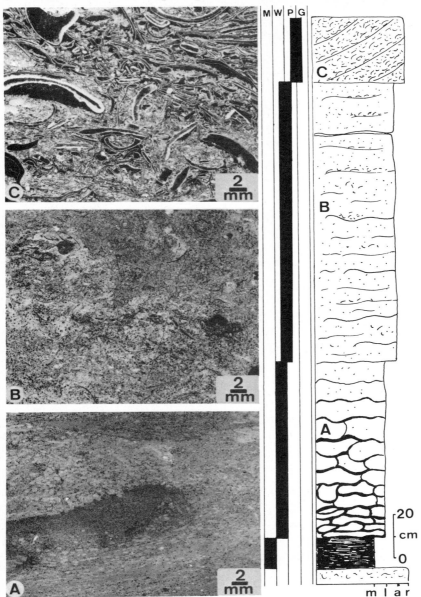

Fig. 56. Detailed log through nodular-to-compact cycle (cf. Fig. 51D). Note upward increase in the abundance and grain size of bioclasts as well as increasing winnowing. Section no. 35, Westheim.

recognizable (Fig. 51D). Their basal parts are rather marly, nodular and strongly bioturbated. Upwards, marl content and the degree of bio-turbation decreases, while bed thickness increases and primary strati-fication is better preserved, producing a more compact weathering appearance. Microfacies change from mudstones and pelletal wackestones (pelmicrites) through calcarenitic into calciruditic skeletal pack-stones at the top (Fig. 56A-D).

Discussion. Marly and strongly bioturbated nodular limestones at the base of these sequences reflect a quiet, sheltered and/or somewhat deeper depositional environment, in which primary stratification was largely obscured by burrowing of Teichichnus. Upward changes in micro-facies and stratification reflect a transition into a higher-energy, shallower setting with small incipient skeletal blankets and shoals.

3.1.6. Thickening-upward cycles (Fig. 51E, 57)

Description. These sequences are generally between 1 and 7 m in thick-ness and are most abundant in more open marine settings.

The lower part is generally a 10-70 cm thick marlstone unit, often with gutter casts (Aigner & Futterer, 1978). Some of the thicker and most widespread marlstone horizons are being used as marker beds in Upper Muschelkalk lithostratigraphy (e.g. "Tonhorizont alpha, beta", etc.), although their cyclic context had not been recognized. The basal marlstone horizon passes upward into interbedded thin limestones and marlstones. Upwards, the percentage of marlstone further decreases, while limestone beds become progressively thicker, their bioclast content higher and coarser-grained (Fig. 57A-C).

The upper part of some sequences includes shallow channel fills (Fig. 57C) and cycle tops are commonly formed by a 20-50 cm thick skeletal packstone (Fig. 57D). These top units are mostly composite beds and consist of a number of amalgamated layers, wedges and channel-like lenses. Reworked pebbles, including black pebbles (Barthel, 1974; Strasser & Davaud, 1983) are abundant and commonly bored and en-crusted. The trace fossil content also changes markedly in these sequences. While their lower part is characterized by Rhizocorallium irregulare and Planolites/Palaeophycus burrows, Teichichnus is commonly present further up. The topmost part of these sequences often displays Glossifungites-type burrows.

THICKENING-UPWARD CYCLE

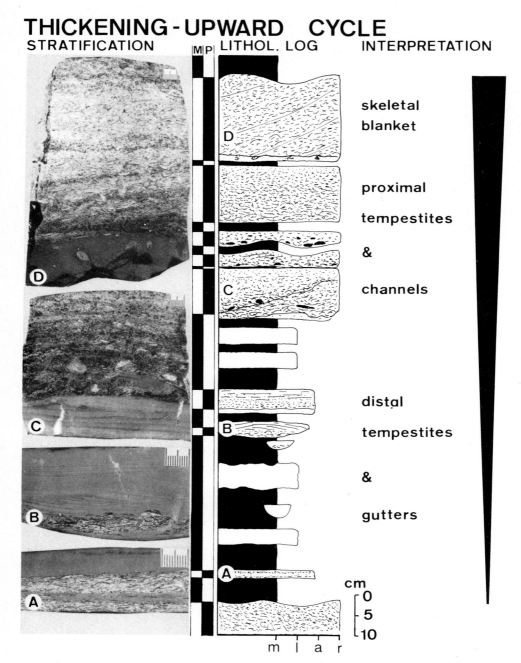

Fig. 57. Detailed log through a thickening-upward cycle (see Fig. 51E). Note upward changes from distal tempestites with gutter casts (A,B), into proximal tempestites (with surge channels, C) and finally into a cross-bedded unit of skeletal packstone (D). Section no. 1, Steinbächle (from Aigner, 1984).

Like the microfacies, stratification and fauna, paleocurrents show marked changes within these sequences. Fig. 58 shows paleocurrent data collected from various parts of one cycle over a wider regional area. Gutter casts in the lower part of the cycle are oriented parallel to the shoreline (cf. Aigner & Futterer, 1978). Sole marks (bounce casts and bipolar prod casts) at the bases of distal tempestites also show longshore but bipolar orientations. Upwards, sole marks show a larger scatter and shift towards a more on-offshore orientation. Channel fills and imbrication in proximal tempestites, however, indicate off-shore sediment transport in the upper part of these cycles.

Discussion. As in the previous sequences, all upward changes indicate shallowing-upward trends. The basal marlstone horizon is interpreted to represent deepest water conditions. The limestone sheets and gutter casts are distal expressions of storm-induced flows (distal tempestites), while upward changes into thicker and coarser-grained limestones mark the transition to proximal tempestites, associated with storm-surge channels. The topmost unit represents a complex amalgamation of a number of events in relatively shallower water, which has in some cases led to the development of skeletal blankets and incipient shoals.

Similar to the lithofacies, the changing ichnofauna reflects a change from deeper-water quiet softgrounds to shallower-water event-agitated firmground surfaces (Seilacher, 1967). The association of these firm-grounds with blackened, bored and encrusted pebbles indicate periods of low net sedimentation and non-deposition. The paleocurrents show a change in the hydraulic regime during shallowing from a dominance of longshore flow in deeper, offshore water (distal) to a dominance of offshore directed sediment transport in shallower, nearshore (proximal) water.

Fig. 58. Detailed log of thickening-upward sequence traced over a regional (Kocher-Jagst) area with paleocurrent data from various parts of the same sequence. In its lower, "distal" part, gutter casts and sole mark directions are largely alongshore, perpendicular to wave ripple crests. Upwards, wave ripples show more interference patterns, sole marks swing into on/offshore direction but are more variable, and imbrication in the uppermost proximal beds indicates offshore directed flow. SHA = Schwäbisch Hall, CR = Crailsheim. Sequence below "Tonhori-zont beta".

→

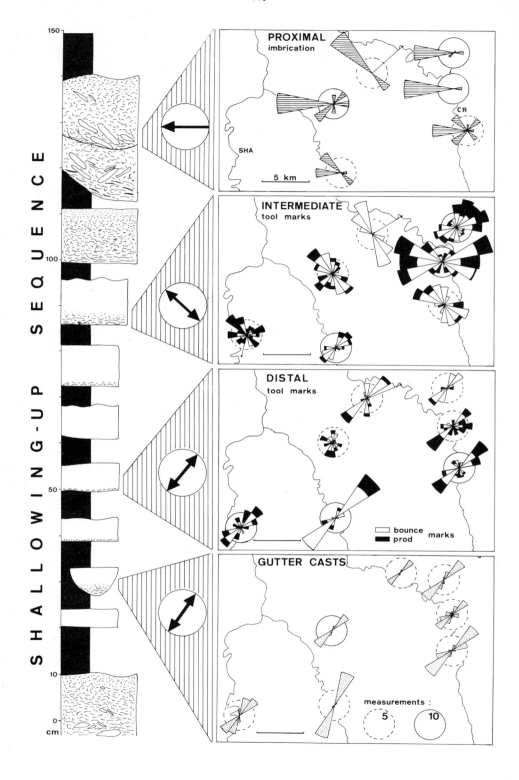

3.1.7. Conclusions: transgressive/regressive dynamics (Fig. 59)

Although the character of vertical facies sequences varies from nearshore to offshore in the Muschelkalk Basin (see previous sections),
some generalisations can be made for an "ideal" cycle (Fig. 59).
Sequences commonly start with thin but widespread marlstone horizons
("Tonhorizonte") many of which are used as lithostratigraphic markers
and comprise such trends as coarsening-upwards, thickening-upwards,
dark-to-light colour changes upwards, and more physical sedimentary
structures upwards. Paleocurrents tend to change upwards from
distinctly longshore orientation to offshore. Faunal associations
reflect changes in substrate conditions from (1) basal softgrounds to
(2) episodic firm- and shellgrounds to (3) either more permanent firm-
and shellgrounds or to mobile carbonate sands at cycle tops.

Although their genetic significance had escaped attention, many cycle
top units have also been used for lithostratigraphic correlation (e.g.
"Mittlere Schalentrümmerbank"). They generally represent winnowed,
storm-amalgamated and -condensed horizons that provided firm- and
shellgrounds. In some instances, these cycle top horizons allowed geographically widespread colonisation by specific fixosessile brachiopods and crinoids. Such "ecologically fingerprinted" units (e.g.
Spiriferina-Bank, Holocrinus-Bank; see Hagdorn, 1985) are especially
powerful marker beds for "event-stratigraphic" correlation.

All patterns of upward change in lithology, paleoecology and
paleocurrents within the above described vertical facies sequences
document shallowing-upward cycles. Shallowing-upward cycles of a
variety of types form a fundamental motif of shallow-water carbonate
accumulation (for summaries see Wilson, 1975; James, 1980; Enos,
1983). Such sequences are so pervasive because carbonates tend to
accumulate at rates much greater than average rates of subsidence can
accommodate (e.g. James, 1980). Consequently, carbonate accumulations
will repeatedly and rapidly build up to sea level, thus generating
vertical repetitions of shallowing-upward cycles.

Since many individual cycles in the present example can be correlated
over large parts of the basin (see below), regional rather than local
mechanisms are likely to be main controls for this cyclicity. In
accordance with Wilson (1975), the Muschelkalk sequences are interpreted as asymmetric transgressive/regressive cycles caused by a rapid
rise in relative sea level, followed by vertical accretion and seaward

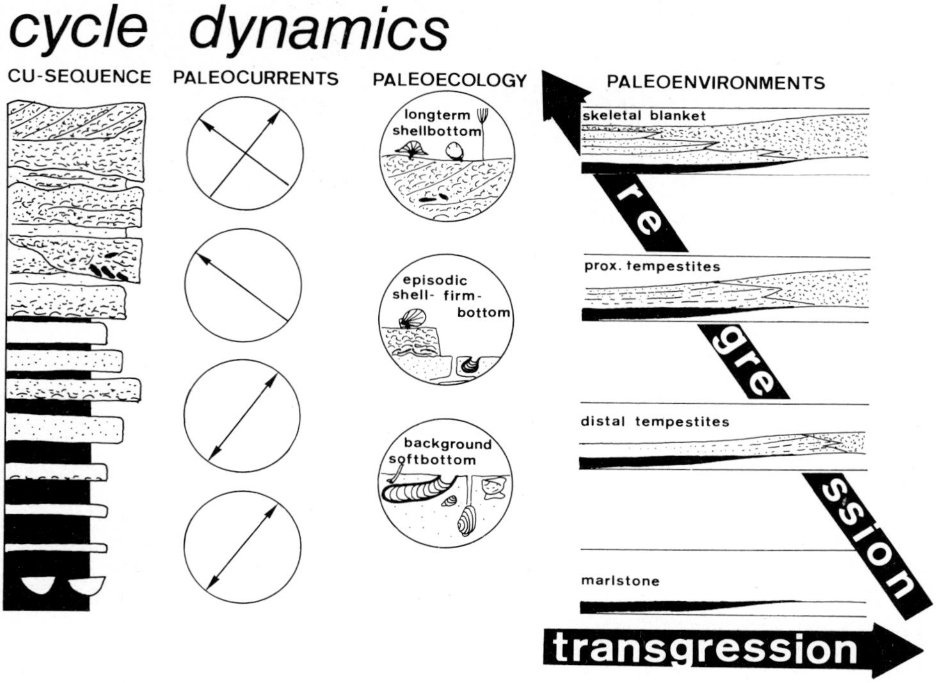

cycle dynamics

Fig. 59. Summary on trends in lithology, paleocurrents and faunas through an "ideal" coarsening-upward cycle and interpretation in terms of transgressive/regressive dynamics. For further explanation see text.

outbuilding of the shoalwater complex. The distribution of these cycles within the entire basin gives some clues regarding a tectonic versus an eustatic prime control for the changes in relative sea level (see below).

Using the rough "rule of thumb" proposed by Matthews (1984), that the sea level rise to explain any given thickness of shallow-marine sediments is approximately 30-40 % of the total sediment thickness observed, only minor changes of relative sea level in the order of one meter or less would be required to produce such cycles.

A rough estimate of possible time scales for these Muschelkalk cycles based on average cycle thickness and the duration of Upper Muschelkalk deposition gives periods of 60,000 - 440,000 years per cycle. Published time estimates for similar cycles are about in the same order of magnitude (generally betweeen 20,000 and 600,000 years, Wilson, 1975). Thus the rate of cycle formation is geologically so rapid, that it mostly falls beyond normal biostratigraphic resolution. For the Upper Muschelkalk, this consideration justifies the use of those transgressive and regressive units which are particularly widespread, as markers both for litho- and for time-stratigraphic correlation.

3.2. Lateral sequences: carbonate ramps

Lateral facies changes from coastal to basinal environments in the Upper Muschelkalk were already recognized by Wagner (1913b), Klein- sorge (1935), Vollrath (1938-1970), Skupin (1970), Schäfer (1973) and Hagdorn (1978, 1982). However, no attempt has yet been made to provide a comprehensive model accounting for depositional dynamics.

Thanks to an abundance of quarry sections, many units of the Upper Muschelkalk can be traced laterally and their facies changes examined. The general facies pattern is that nearshore skeletal and oolitic sediment bodies pass offshore into bioclastic sheets interbedded with fine-grained sediments and finally into interbedded lime mudstones and marlstones. Since shallow-water sediments grade without break into deeper-water deposits and since significant slump deposits are generally absent, a gently sloping carbonate ramp can be inferred (Ahr, 1973; Read, 1982 a,b). The present-day Persian Gulf provides a good actualistic model (Purser, 1973; Wilson & Jordan, 1983) for carbonate ramps (Fig. 60A).

In the lower Upper Muschelkalk, shallow-water skeletal sediment bodies are dominated by crinoid ossicles (crinoidal ramps), while those in the upper Upper Muschelkalk comprise only oolitic and shelly material (shelly/oolitic ramps, see Fig. 60). The present examples can be regarded as "ramps with barrier bank complexes" and "ramps with barrier ooid complexes" using Read's (1982 b) classification.

3.2.1. crinoidal ramps (Fig.61)

Description. The general lateral facies succession in the lower Upper Muschelkalk ("Trochitenkalk") is that nearshore massive and partly oolitic crinoidal limestones pass offshore into shelly limestones and marlstones in the basin center (Fig. 60B). Facies types grade conti-

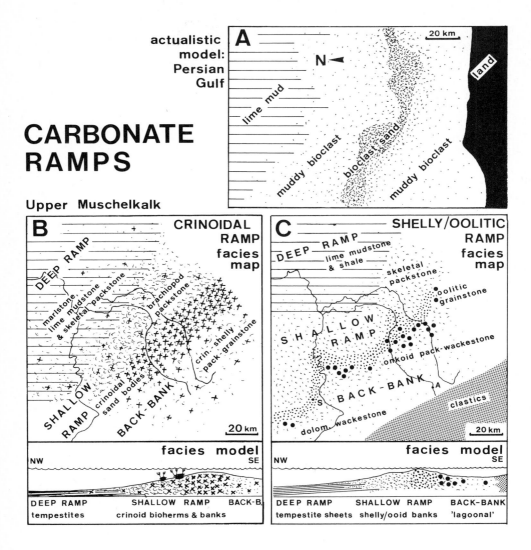

CARBONATE RAMPS

Upper Muschelkalk

Fig. 60. A) Modern example of a carbonate ramp from the Persian Gulf (simplified after Wilson & Jordan, 1983). B) Generalized facies distribution on Upper Muschelkalk crinoidal ramp ("Trochitenbank 4" cycle, based on a compilation of data from Vollrath, 1957, 1958; Hagdorn & Simon, 1981, Skupin, 1970, Hagdorn, 1978, and personal observations). C) Generalized facies distribution on Upper Muschelkalk shelly/oolitic ramp (upper part of "Region der Schalentrümmerkalke", based on a compilation of data from Wagner, 1913b; Vollrath, 1955a; Schröder, 1967; Hagdorn & Simon, 1981; and personal observations). A = Aalen, S = Stuttgart. Black dots are oncolites. Note the similarity between actualistic model and Muschelkalk ramps both in facies successions and in dimensions (all maps drawn to the same scale).

nuously and without abrupt change into each other. To depict details
of lateral changes, Fig. 61 shows one 1 - 6 m thick shallowing-up
cycle ("Trochitenbank 4" cycle) across a nearshore-offshore transect.
Seaward of a zone of unexposed dolomitic crinoidal limestones (only
recorded in cores, see Hagdorn & Simon, 1981), a prominent belt of
massive crinoidal limestones ("Crailsheimer Riffkalk", Vollrath, 1957,
1958) occurs. This complex consists of crinoid-mollusc-brachiopod
pack- to grainstone and includes zones of oolitic grainstone. Small
pelecypod-crinoid bioherms also occur within this complex (Hagdorn,
1978). Further offshore, the crinoid content decreases and brachiopod
packstone is dominant. Still further into the basin, fine-grained
mollusc-brachiopod packstone interbedded with lime mudstone and marlstone predominates.

Discussion. Dolomitic crinoidal limestones in the very marginal parts
of the Upper Muschelkalk basin probably represent a "lagoonal" depositional environment, protected by a barrier-like belt of crinoidal
and partly oolitic banks and shoals. Many of the crinoidal ossicles in
these nearshore accumulations seem to be derived from pelecypod/
crinoid bioherms that occur in more seaward parts of the crinoidal
shoalwater complex. The dynamics of crinoidal bank accumulation may be
similar to the buildup of modern nearshore skeletal banks, such as the
Safety Valve in South Florida (see part I). During episodic storms,
wave stirring and landward oriented wind drift currents would be able
to pile up skeletal debris and construct nearshore skeletal banks
(Fig. 62). Landward transport of crinoid ossicles and accumulation
into nearshore crinoidal sand was also suggested by Ruhrmann (1971)
for Paleozoic crinoidal limestones. He assumed that the low bulk
density of crinoid particles would facilitate transport in suspension.
In spite of some landward transport, possibly in form of spillover
lobes, sediment accumulation appears to be largely in situ and parautochthonous (Hagdorn, 1978), simlar to possible actualistic counterparts
(see part I).

Fig. 61. Transect through a crinoidal ramp: nearshore-offshore
variation of one shallowing-upward cycle in the lower part of the
Upper Muschelkalk (NW-SE section through map in Fig. 60B; "Trochitenbank 4 " cycle in Fig. 71). Lower part: variation in field appearance
of 1 (F) to 6 m (H) thick shallowing-upward cycle; note wedging out of
massive crinoidal bank facies at cycle top and increase in marlstone
in lower part of cycle towards the basin (quarries: F = Asbach, G =
Garnberg, H = Neidenfels). Midde part: microfacies variation in cycle
top unit ("Trochitenbank 4"). Width of fotos 2.5 cm. Upper part:
schematic facies model for crinoidal ramp. Crinoidal bioherms not
included (see Hagdorn, 1978).

CRINOIDAL RAMP

CRINOIDAL RAMP DYNAMICS

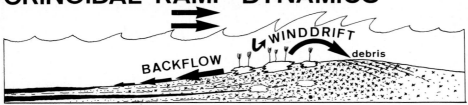

Fig. 62. Model for the storm-dominated dynamic regime on crinoidal ramps. Onshore wind-drift currents in surface water pile crinoid debris (largely from crinoidal bioherms) into shallow-ramp skeletal banks, possibly similar to storm-generated accretion of nearshore carbonate banks in South-Florida (see part I). Nearshore water build-up is compensated by offshore directed bottom return flows that cause tempestites in the deep ramp. Proximal tempestites include crinoid ossicles imported from the nearshore, in contrast to distal ones, that are dominated by reworked but parautochthonous shelly faunas.

The general decrease in crinoid ossicles from proximal to distal has ecological as well as sedimentological reasons. In the Muschelkalk Sea, crinoids seem to have lived mainly in shallower, marginal parts of the basin (Hagdorn, 1978, 1985). Offshore decreasing crinoid content also reflects the decreasing capacity of offshore directed storm-generated flows to transport particles from shallower to deeper parts of the basin. Brachiopod-dominated limestones seaward of the crinoid bank complex are interpreted as somewhat deeper environments with brachiopod colonies related to small bioherms (see Hagdorn, 1978, Fig. 22).

Principally similar lateral facies sequences that may be interpreted as "crinoidal ramps" have been recorded by Laporte (1969) from the Devonian Helderberg Formation and by Wilson (1975) from the Mississippian of the Williston Basin.

3.2.2. Shelly/oolitic ramps (Fig. 63)

Description. In the upper part of the Upper Muschelkalk, the nearshore shoalwater complex does not include crinoid ossicles but is largely composed of shelly and oolitic sediment.

Fig. 60C shows the regional distribution and overall zonation of major
facies types in an approximately 2 - 5 m thick shallowing-upward cycle
("Kornstein I cycle", cf. Fig. 74) based on a compilation of published
data and personal observations. Following a zone of coastal clastics,
an area of partly dolomitized rocks with scattered oncolites can be
recognized. Oncolites are most abundant and well-developed in a
distinct zone landward of a narrow belt of oolitic grainstones, that
in turn pass seaward into pack- to grainstones. These grade laterally
into interbedded thin skeletal sheets, lime mudstones and marlstones
toward the basin center.

Fig. 63 shows the variation in microfacies and sequences across the
area represented in Fig. 60C. Strongly bioturbated and unsorted fine-
grained skeletal wacke- to packstones with coated grains and black
pebbles at the basin margin pass into coarser-grained oncolitic pack-
to grainstones. These give way to a zone of well-sorted oolitic
grainstones, that pass seaward into well-sorted skeletal packstone
with abundant micritic envelopes. Still further offshore, sorting and
the abundance of micritic envelopes as well as grain size of particles
generally decrease while the mud content increases. In the most
offshore section, pellet-rich calcarenites, and thin fining-up layers
predominate.

Discussion. The facies belt with thick cross-bedded oolitic and shelly
pack- to grainstones is interpreted as shallow ramp shoalwater complex
similar to Holocene carbonate sand bodies from modern ramps such as
the Persian Gulf (Loreau & Purser, 1973, Wilson & Jordan, 1983). The
zone of most prominent oncolite development landward of the oolite
banks is interpreted as shallow ramp to lagoonal facies. The partly

Fig. 63. Transect through a shelly/oolitic ramp: nearshore-offshore
variation of one shallowing-upward cycle in the upper part of the
Upper Muschelkalk (NW-SE section through map in Fig. 60C; "Kornstein
I" cycle in Fig. 74). Lower part: lithologic variation of cycle; note
that shallow ramp shoalwater complex wedges out basinward (towards the
left) and gets replaced by thin-bedded limestone/marlstone alternations. Middle part: microfacies variation in cycle top unit ("Kornstein I"); note change from bioturbated shell hash with small black
pebbles (F, back-bank) to ruditic grainstone with oncoids (E) to
oolitic grainstone (D), to well-sorted then poorer sorted shelly packstone (C and B), to laminated calcarenite (A). Width of fotos 2.5
cm.Upper part: schematic facies model for shelly/oolitic ramp. ⟶

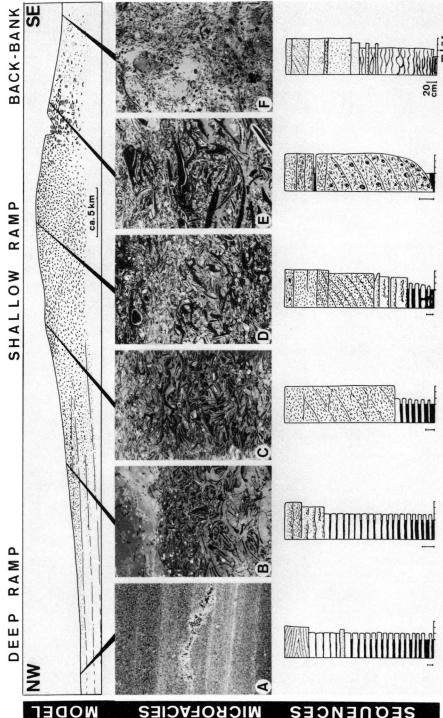

dolomitized wacke- to packstones between the oncolite zone and the coastal clastics seem to represent a protected to lagoonal environment behind the barrier bank/shoalwater complex (cf. Hagdorn & Simon, 1981). Seaward of the oolite bank facies, shelly packstones are interpreted as shoals and blankets of mollusc sand in slightly deeper water than the constantly agitated ooid sand banks. Decreasing sorting and increasing mud content indicates a transitional environment between the shallow ramp and deep ramp. The latter is represented by a dominance of tempestites interbedded with lime mudstones and marlstones. This lateral facies zonation is remarkably similar to modern ramps in the Persian Gulf (Wilson & Jordan, 1983) and to some ramps described from the fossil record (e.g. Markello & Read, 1981, 1982).

3.3. Paleoecological trends (Fig. 64)

3.3.1. Ramp biofacies

Description. Systematic changes in the composition of benthic associaions from nearshore to offshore in the Upper Muschelkalk have been documented by Hagdorn (1978, 1982). He recognized the following basic pattern (Fig. 64) in the upper Upper Muschelkalk:

(1) Rapidly burrowing deposit feeeders (Myophoria, Trigonodus) and byssate suspension feeders (Bakevellia, Entolium) in mobile substrates of carbonate sand bodies.

(2) Epifaunal shell-, firm- and hardground assemblages (Placunopsis, Pleuronectites, Plagiostoma, Coenothyris) in an intermediate zone;

(3) a dominance of infaunal softbottom assemblages (mostly deposit feeders such as Myophoria, Paleonucula, Entalis) and byssate recliners (Hoernesia) further offshore.

Discussion. In modern shelves, wave base and storm wave base mark distinct boundaries between animal communities (Dörjes & Hertweck, 1975). The Muschelkalk trends are also understandable in the context of a storm-dominated ramp setting with decreasing effects of storms and changing substrate conditions toward the offshore (cf. Hagdorn, 1982):

(1) <u>shallow ramp</u> carbonate sand bodies were largely above wave base and thus constantly agitated. Consequently these areas could only be colonized by rapid burrowers or flexibly attached epifauna.

(2) The <u>transitional</u> zone between carbonate sand bodies and the deep ramp probably represents the zone between "fair-weather" wave base and the wave base of average storms. Here, frequent storm reworking and scouring lead to proximal tempestites with shelly, firm or hard surfaces that provide opportunities for byssate or fixosessile epifaunal suspension feeders.

(3) <u>Deep ramp</u> areas were probably only reached during exceptional storms. These areas are characterized by mostly muddy substrates except for rare storm-introduced coarser sediment. The muddy background sediments were therefore inhabited by shallow infaunal deposit feeders with some deeply burrowing suspension feeders. Epifaunal associations are largely restricted to post-event colonisation, i.e. to tempestite tops.

3.3.2. Ramp ichnofacies

<u>Description</u>. A comprehensive zonation of trace fossils across bathymetric gradients was developed by Seilacher (1967). His scheme was applied by numerous workers and has proved to be a valuable tool in facies analysis. While Seilacher´s scheme provides a general subdivision into continental to deep sea ichnofacies, zonations across shallow shelf environments have been refined by later workers.

Across Muschelkalk carbonate ramps, trace fossil associations show marked trends, similar to zonations in the Jurassic (e.g. Farrow, 1966; Ager & Wallace, 1970; Fürsich, 1975). In this context, only the most abundant and most characteristic lebensspuren are considered (Fig. 64):

(1) <u>Protected to lagoonal</u> environments are mostly thoroughly reworked, probably by mobile deposit feeders that left few distinct traces, except for <u>Teichichnus</u>.

(2) In the <u>shallow ramp</u> shoalwater complex, trace fossils are relatively rare. Occasional hardgrounds are penetrated by small borings such as <u>Calciroda kraichgoviae</u> (Mayer, 1952; Voigt, 1975).

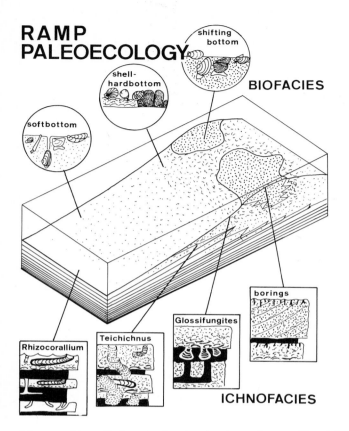

RAMP PALEOECOLOGY

BIOFACIES

ICHNOFACIES

Fig. 64. Paleoecological trends reflecting changes in substrate properties and in energy levels across an "ideal" carbonate ramp. Biofacies change from shallow ramp "shifting bottom" to "shell- and hardbottom" to deep ramp "softbottom" assemblages (after Hagdorn, 1982). Ichnocoenoses in the shallow ramp are characterized by borings and burrows of suspension feeders such as Glossifungites and Thalassinoides , but give way to a dominance of sediment feeders such as Teichichnus and Rhizocorallium assemblages in the deep ramp.

(3) In the transitional areas between shallow and deep ramp, scoured surfaces display Glossifungites (= Rhizocorallium jenense sensu Fürsich, 1974) and Thalassinoides burrows (mostly larger types, 1-3 cm diameter). Otherwise these areas are characterized by abundant Teichichnus spreiten with minor Rhizocorallium irregulare (sensu Fürsich, 1974) and Planolites/Palaeophycus (sensu Pemberton & Frey, 1982).

(4) Limestone/marlstone alternations of the deep ramp are dominated by Rhizocorallium irregulare. Bedding planes are often covered with

different types of "stick burrows", some of which are of the Planolites type (with active backfilling structures), others of the Palaeophycus type (without backfilling, see Pemberton & Frey, 1982). Balanoglossites type burrows (cf. Kazmierczak & Pszczolkowski, 1969) also occur on the tops of lime mudstones. Other less frequent trace fossils are smaller (0.5-2 cm) Thalassinoides, Granularia and Phycodes.

Discussion. Much like the biofacies, these trends reflect changing substrate properties and energy levels across the ramp: a dominance of feeding structures (Rhizocorallium, Planolites, Phycodes) in deep ramp settings indicates relatively quiet softgrounds during background sedimentation. Following episodic erosional and depositional events, however, dwelling burrows (e.g. Glossifungites, Balanoglossites) occur as post-event faunas on firmer substrates within the deep ramp environments. An increase in dwelling structures toward the shallow ramp reflects an increasing availability of storm-scoured firmgrounds and of storm-reworked shelly substrates in shallower water.

3.4. Paleocurrents

In the following, paleocurrent data collected from about 60 localities are presented in a series of maps. Most of the data are from the upper part of the Upper Muschelkalk and therefore refer mainly to the shelly/oolitic type of carbonate ramp.

3.4.1. Wave ripples (Fig. 65)

Wave ripple fields on "proximal", thicker skeletal sheets often form the bases of quarry operations and are therefore commonly well-exposed and well accessible for measurement. Stratigraphically, most measurements of Fig. 65 are derived from around the spinosus-zone, although some data from other parts of the sequence are also included.

Generally, wave ripple crests in more central parts of the Muschelkalk Basin tend to be oriented almost perpendicular to the shoreline. Towards more marginal areas, however, wave ripple crests tend to "swing around" into orientations oblique and almost parallel to the

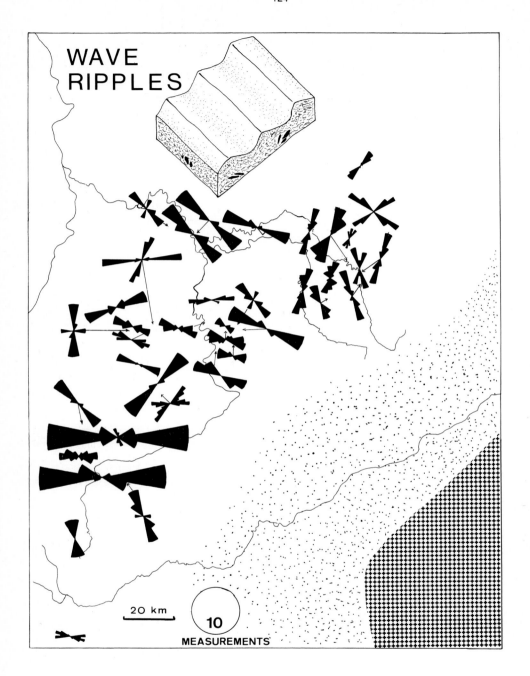

Fig. 65. Wave-ripple orientation (mostly from around the spinosus-zone) indicating longshore oscillatory flow.

shoreline. This pattern would seem to indicate dominantly longshore winds and storm tracks from the Tethys to the NE, which would generate dominantly alongshore wave trains in offshore waters. Due to the Coriolis force, however, alongshore blowing winds would drive surface water eastward (landward) to the right of the wind, thus accounting for nearshore wave ripple orientation oblique or parallel to the coast-line. Wind and storm tracks from the Tethys toward the NE can also be predicted using the paleostorm model of Marsaglia & Klein (1983; see Fig. 29A). Winds from the South in the Middle Triassic have also been recorded in the Northern Alps (Zankl, pers. comm.) and in the Dolomites (Blendinger, 1985).

3.4.2. Cross-bedding (Fig. 66)

As indicated above, it is often very difficult to recognize and measure cross-bedding from freshly cut rock surfaces in quarries; weathered sections generally give a much better picture.

Fig. 66 shows cross-bedding measurements mostly taken from massive oolitic or skeletal pack- to grainstones of the "Region der Schalen-trümmerbänke" and the "Region der Oolithbänke"; few measurements from other stratigraphic intervals of the Upper Muschelkalk were also in-cluded since they were consistent with the other data. Clearly, most foresets dip alongshore towards the NE, few onshore (E,SE), or off-shore (NW,W). This indicates mostly alongshore energy flux and sand body migration.

As pointed out by several authors (e.g. Johnson, 1978), cross-strati-fication patterns in shallow-marine sands are often difficult to assign to a wind-driven versus a tidally driven circulation. The uni-modality, general context and relationship to other paleocurrent indicators, however, favours a dominance of wind-driven currents. The relationship between foreset direction and wave-ripple orientation is especially suggestive of a wind/storm-driven hydraulic regime. Longshore orientation of foresets would thus record sand body migration in direct response to longshore winds and storm tracks, while onshore foresets in marginal parts would represent the response to nearshore water set-up during storm events.

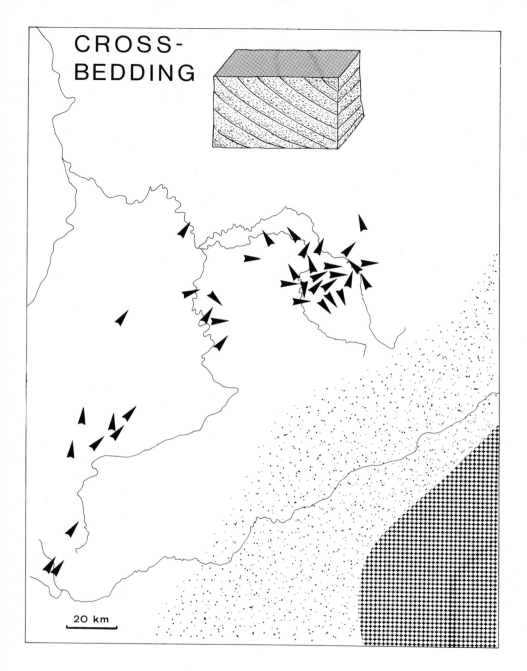

Fig. 66. Cross-stratification measurements from massive shelly and oolitic limestones indicating dominantly longshore (northeastward) sand body migration, but also onshore and some offshore transport.

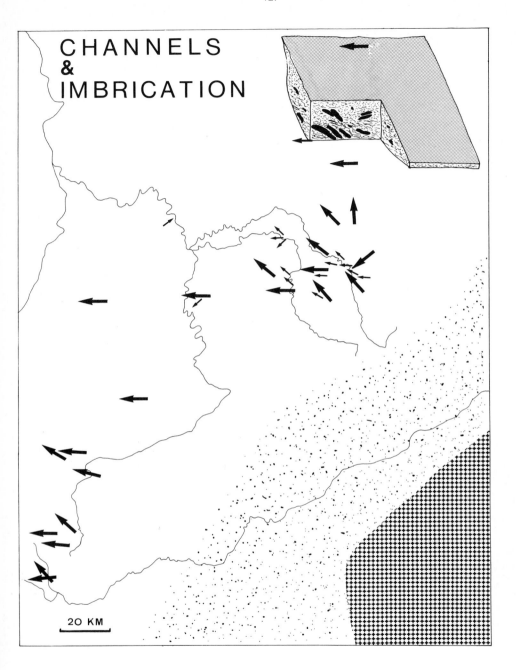

Fig. 67. Orientation of channel fills and imbrication of intraclasts indicating offshore directed flows.

3.4.3. Surge channels and imbrication (Fig. 67)

Relatively small (1-30 m long, - 1 m deep) skeletal channel fills (see section 2.7.) have been found in a number of quarry sections. Channel orientation is always onshore-offshore (E-W; NW-SE), more or less perpendicular to the shoreline. Imbrication of pebbles always indicates offshore oriented flow and sediment transport.

In the context of storm depositional systems, these channels are inter-preted as being caused by offshore flowing water masses (backflow of storm surge/rip current/gradient current type) that compensate for nearshore water set-up during storms.

3.4.4. Gutter casts (Figs. 68, 69)

Gutter casts are sediment fills of small but often meter-long erosional runnels (see Fig. 68, details see Aigner & Futterer, 1978). They occur within the offshore thin-bedded limestone/marlstone alter-nations and their fill-sediment shows sequences analogous to distal tempestites (Fig. 68D). In contrast to Duringer (1982) it is main-tained that these runnels are not indicative of subaerial exposure but were clearly formed subaqueously. This can be inferred from (1) their stratigraphic context, (2) the near absence of branching, strong meandering and lateral migration typical for tidal channels, and (3) their distinct NE-SW orientation throughout the South-German Basin (Fig. 69). The NE-SW orientation corresponds to the dominant paleo-currents expressed by bounce and prod marks on the soles of distal tempestites with which the gutter casts are interbedded (cf. Fig. 43A).

Although the physics of gutter cast formation is not yet fully under-stood (see e.g. Whitacker, 1973; Aigner & Futterer, 1978; Allen, 1982), the participation of <u>oscillatory</u> flow components is indicated by two factors: (1) prod marks at the basal erosion surfaces of gutter casts are commonly bipolar, in agreement with observations made by Bloos (1982) in Liassic gutter casts. (2) Wave ripples at the top of gutter casts are commonly perpendicular or slightly oblique to the axis of the gutter (Fig. 68C).

However, internally structures, such as imbrication and climbing ripple lamination (Fig. 68E) indicate superimposed <u>unidirectional</u>

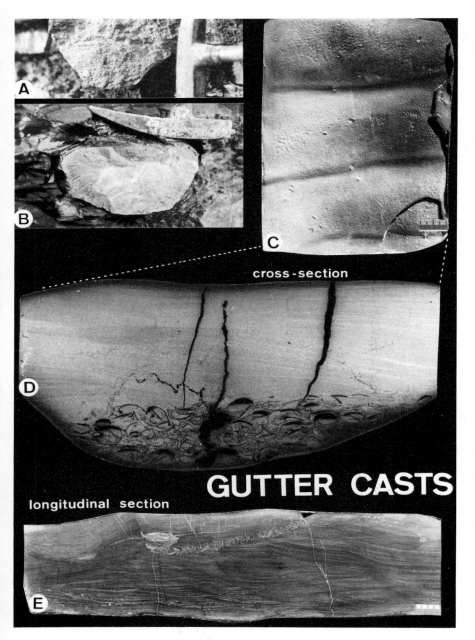

Fig. 68. Some characteristics of gutter casts. A) - B) Field
appearance either at base of tempestites (A), or isolated gutter cast
within marlstone (B). C) Wave ripples at gutter cast top, ripple
crests perpendicular to axis of gutter. D) Cross-section through
gutter cast, note similarity to tempestite sequence. E) Axial (longi-
tudinal) section through gutter cast with climbing ripple lamination
and Rhizocorallium-spreite from top.

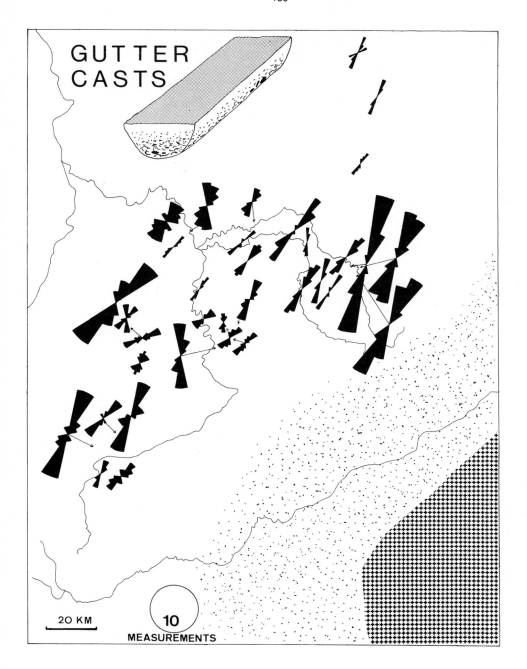

Fig. 69. Orientation of gutter casts(from   around   the   spinosus-zone)
indicating longshore flow.

(alongshore) flow components during the fill of the gutter (14
sectioned gutter casts showed northeastward, 8 southwestward flow).
Gutter cast formation may therefore be the product of <u>combined flows</u>,
with superimposed oscillatory and unidirectional components as they
are typical for storm flows (e.g. Swift et al., 1983). Again, they
would indicate alongshore storm tracks from the Tethys to the NE and
support the hydraulic pattern indicated by other paleocurrents
(above). In addition, longshore SW-NE flows are also indicated by the
orientation of fossils on bedding planes as reported by Mayer (1955)
and by Seilacher (1960) .

3.5. Conclusions: carbonate ramp dynamics (Fig. 70)

The aim of this chapter has been to integrate (1) vertical rock
sequences with (2) lateral facies patterns, (3) paleoecological trends
and (4) paleocurrent data in order to reconstruct the dynamic
processes on Muschelkalk carbonate ramps. These reconstructions refer
principally to the shelly/oolitic type of ramps in the Upper Muschelkalk.

The dynamics of ramp systems appear to differ markedly from rimmed
carbonate shelves in that ramps lack protective reef belts. This makes
ramps extremely susceptible for the effects of swell, waves and
storms. It is therefore understandable that both the details of the
stratification and the overall facies organisation are largely
controlled by episodic high-energy events such as storms. The
relatively uniform and gently sloping sea-floor topography of such
ramps seem to have favoured the predominance and wide distribution of
relatively thin depositional units. Marked nearshore-offshore changes
in the composition of benthic asociations are also understandable in
the context of a storm-dominated ramp system.

Although the shallow ramp oolitic sediment bodies were almost
constantly agitated, major changes (e.g. formation of spillover lobes,
cutting of channels) only take place episodically during storms.
Relatively low-energy day-to-day conditions in the lagoonal area
behind the barrier complexes were only episodically interrupted by
high-energy events. In the transitional zone between shallow and deep
ramp, thick and often composite units (amalgamation of several
fining-up units) represent proximal tempestites. Storm surge channels

funneled sediment toward the offshore. The deep ramp and basinal lime-stone/marlstone alternations represent autochthonous background sedimentation interrupted by distal tempestites and gutter cast erosion.

An integration of facies sequences and paleocurrent data allows to reconstruct in detail the hydraulic regime on the present carbonate ramps (Fig. 70). Close parallels to alongshore storm flows documented by Swift et al. (1983) from the Atlantic Shelf of North America are apparent, that contrast against the largely onshore storms in the German Bight (see part I). Storm tracks in the Muschelkalk are likey to have been alongshore from the Tethys in the South into the Germanic Basin towards the NE. This is inferred from the orientation of wave

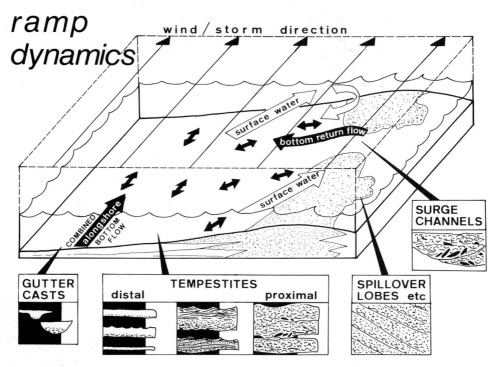

Fig. 70. Model for the storm-dominated hydrodynamic regime on Upper Muschelkalk carbonate ramps, in analogy to actualistic models (Swift et al., 1983). Alongshore winds and storms (from the Tethys in the SW to the NE) induce combined geostrophic bottom flows (gutter casts and distal tempestites), but the Coriolis effect forces surface water to the right (landward). This leads to nearshore water build-up (formation of spillovers) which is compensated by offshore directed bottom return flows (erosion of surge channels and deposition of proximal tempestites).

ripples, bipolar tool marks, gutter casts and cross-bedding foresets. Wind and storm tracks from the South are also predicted using the paleostorm model of Marsaglia & Klein (1983), and agree with wind patterns recorded from the Alpine Middle Triassic.

Alongshore storm tracks would induce a combined geostrophic bottom flow with alongshore direction just as in the North Atlantic (cf. Swift et al, 1983). Such combined oscillatory/unidirectional flows would explain the bipolar impacts (= oscillatory component) at the base of distal tempestites and are most probably also responsible for gutter cast erosion (Fig. 70). However, wind stress acting northeastward along the Muschelkalk sea would at the same time drive surface water landward, to the right of the wind, caused by the Ekman transport (Coriolis effect). Onshore directed surface flows would set up onshore wind drift currents in shallow water that would lead to the accumulation of skeletal banks, in a way similar to modern examples from South Florida (see part I). Landward transport of surface water is compensated by offshore bottom return flows, similar to gradient currents in the North Sea (see part I) or to coastal downwelling in the Atlantic Shelf (Swift et al., 1983). Such offshore bottom return flows are documented by on-offshore oriented storm surge channels as well as by sole marks and imbrication in proximal tempestites.

Based on their work in the Atlantic Shelf, Swift et al. (1983) speculated that combined flow currents would be the typical response of ancient epicontinental seas to storms. Epicontinental carbonate ramps of the Upper Muschelkalk strongly support this inference and provide the first ancient example of a storm depositional system that allows detailed reconstructions of its hydraulic regime (Fig. 70).

4. BASIN ORGANISATION

4.1. Distribution of minor cycles

In order to document the geometrical arrangement and packaging of minor cycles over larger areas (cf. Fig. 30), logs are assembled in several nearshore-offshore transects through the South-German Basin.

4.1.1. Lower part of Upper Muschelkalk (mol)

The well-established lithostratigraphic subdivision of the lower part of the Upper Muschelkalk is based mainly on an alternation of shelly or crinoidal limestones ("Schalentrümmer-" or "Trochitenbänke") with lime mudstones and nodular lime mudstones ("Blaukalke" and "Brockelkalke"; see Aldinger, 1928; Vollrath, 1955a,b, 1957, 1958; Wirth, 1957; Skupin, 1969, 1970). Previous workers have named and numbered each of these units (e.g. "Trochitenbank 4", "Brockelkalk 6"), showing that they can be used as marker beds correlatable over larger areas.

Detailed re-logging of sections along two transects (cf. Figs. 71, 72) has revealed a marked sedimentary cyclicity not previously noted. All sequences consist of vertically stacked asymmetrical 1-7 m thick cycles of various types (e.g. crinoidal bank cycles, skeletal bank cycles, thickening-upward cycles etc., see section 3.1.). These minor cycles have in common, however, that they start with a marlstone ("Mergelschiefer"), a lime mudstone ("Blaukalk") or a nodular lime mudstone ("Brockelkalk") unit and terminate with a crinoidal ("Trochitenbank") or shelly layer ("Schalentrümmerbank"). Towards the basin margin, some of the crinoidal limestones become thicker and more prominently developed (Fig. 72). Most of the cycles can be correlated over many ten's of km, and some over most parts of the basin. In the transect of Fig. 72 however, correlation becomes difficult between the more marginal and the more basinal sections.

In summary, the entire part of the sequence consists of successive coarsening-upward cycles, each of which shows a shallowing-upwards trend and represents a minor transgressive/regressive pulse.

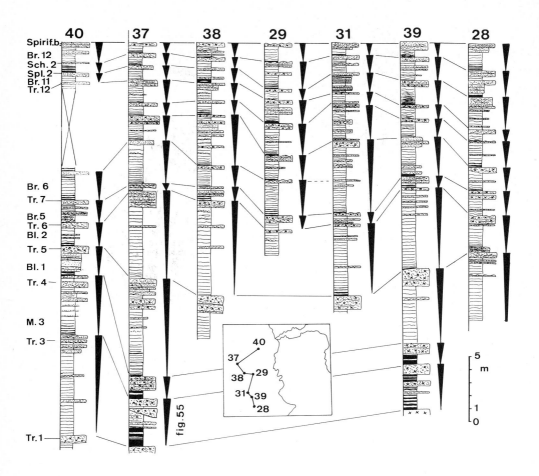

Fig. 71. Log transect through lower part of the Upper Muschelkalk ("Trochitenkalk") slightly oblique to depositional strike, showing correlation of vertically stacked coarsening-upward sequences (black arrows). Lithostratigraphic nomenclature according to Skupin (1969, 1970): Tr. = "Trochitenbank" (crinoidal limestone bed), M. = "Mergel-schiefer" (marlstone), Bl. = "Blaukalk" (lime mudstone), Br. = "Brockelkalk" (nodular lime mudstone), Spl. = "Splitterkalk" (splinter-ing limestone), Sch. = "Schalentrümmerbank" (shelly limestone), Spirif.b. = Spiriferina-Bank (marker bed). For location see Fig. 30, legend see Fig. 72.

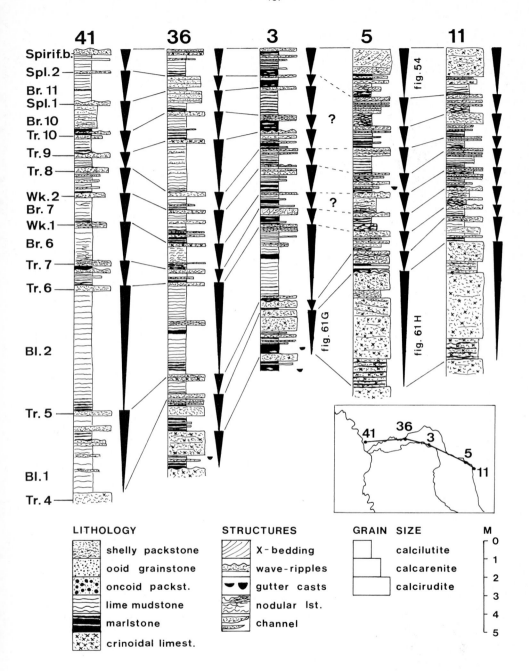

Fig. 72. Nearshore-offshore transect through lower part of the Upper Muschelkalk ("Trochitenkalk"). Note that coarsening-upward cycles can be well correlated in shallower (logs no. 11,5) and in deeper areas (logs no. 3, 30, 41), but that correlation is somewhat difficult between the two zones. Lithostratigraphic nomenclature following Skupin (1969, 1970).

4.1.2. Upper part of Upper Muschelkalk (mo2/3)

Similar to the lower part of the section (see above), some widely distributed marlstone horizons ("Tonhorizonte") and massive shelly/ oolitic units ("Schalentrümmer/Oolithbänke") have so far been used

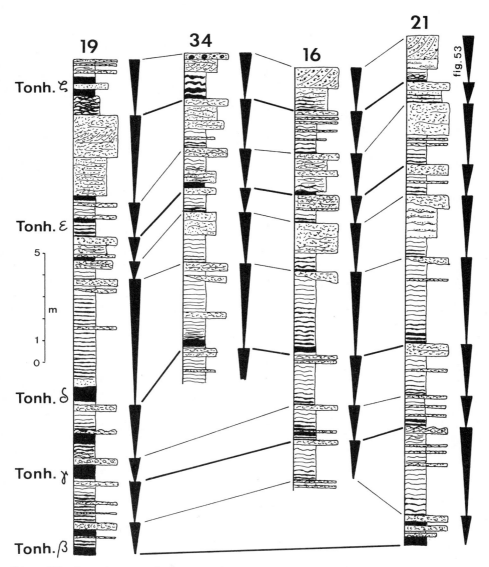

Fig. 73. Log transect in marginal facies province of the upper part of the Upper Muschelkalk showing excellent crrelation of vertically stacked coarsening-upward cycles. The base of many cycles is formed by thin but widespread marlstone horizons (Tonh. = "Tonhorizont") that are being used as lithostratigraphic markers. For location of transect see Fig. 30.

descriptively as lithostratigraphic markers. Detailed re-logging of sections along several transects (Figs. 73-76) revealed, however, that marlstone horizons form the bases, and the shelly/oolitic units the tops of successive asymmetrical coarsening-upward cycles; hence the whole of the Upper Muschelkalk sequence consists of a stack of coarsening-upward (shallowing-upward) cycles. These cycles can also be identified in wells, since the marlstone units show up in gamma-ray logs (Brunner & Simon, 1985).

Laterally, these coarsening-upward cycles can with little difficulty be correlated over many ten's of km (Aigner, 1984). Some cycles, especially those starting with one of the most prominent and named marlstone horizons, can even be traced over much of the basin. Corre-

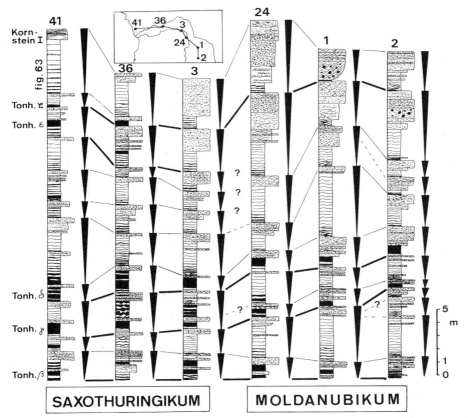

Fig. 74. Log transect through upper part of the Upper Muschelkalk from marginal to basinal areas. Note that coarsening-upward cycles can be well correlated in the marginal (on Moldanubikum) and in the more basinal zone (on Saxothuringikum), but correlation is less certain between these two paleotectonic units. For location of transect see Fig. 30.

lation is least problematic in more marginal parts of the Muschelkalk (see Fig. 73). In transects from marginal to more central parts of the basin, however, correlation of many cycles becomes difficult (Fig. 74), and sometimes almost impossible (Figs. 75,76) along a distinct line because of changes .in the cycle patterns. Moreover, in deep basinal areas of Fig. 76 (logs no. 25, 26), many cycles become totally inverted: they are dominated by fining-upwards, thinning-upwards trends.

The line to which the majority of cycles can be readily correlated corresponds fairly well to the boundary between two underlying Variscan blocks (Moldanubikum and Saxothuringikum, see Fig. 77A). Such a relationship to underlying paleotectonic structures is also apparent in the lower Upper Muschelkalk (see section 4.1.1., Fig. 72), although not as pronounced as in this part of the sequence.

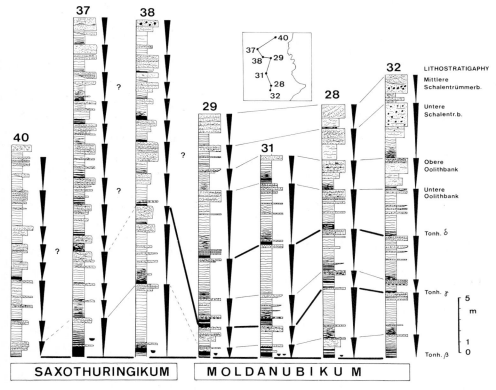

Fig. 75. Log transect through upper part of Upper Muschelkalk. Although coarsening-upward cycles show excellent correlation on the Moldanubian zone, cycle patterns become very different on the Saxothuringian zone which makes correlation very difficult. For location of transect see Fig. 30.

141

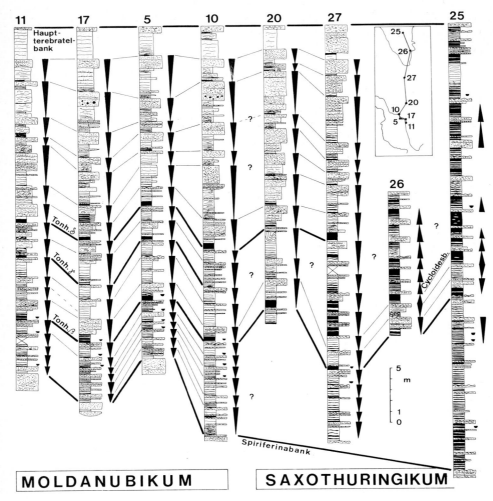

MOLDANUBIKUM **SAXOTHURINGIKUM**

Fig. 76. Nearshore-offshore transect between the well-developed marker
beds of the "Haupterebratelbank" and the "Spiriferina-Bank". Correlation of coarsening-upward cycles on the Moldanubikum is excellent,
but very difficult with the logs on the Saxothuringikum. Note inverted, fining- and thinning-upward cycles are present in the two deep
basinal sections no. 25 and 26. For location of transect see Fig. 30.

4.1.3. Discussion

Vertical stacking of shallowing-upward cycles is a very common motif
of shallow-marine carbonate sequences (e.g. Wilson, 1975; James, 1980,
Goodwin & Anderson, 1980a,b; Enos, 1983). Such sedimentary cycles are
widely used as a tool for chronostratigraphic correlation, because of
their finer resolution compared to the evolution of most organisms

upon which biostratigraphy is based (e.g. Wilson, 1975; Somerville, 1979; Busch & Rollins, 1984). This method has also been called "kick-back" correlation (Irwin, 1965; Matthews, 1984); "event correlation" (Ager, 1981; Dixon et al., 1981; Cant, 1984) is a similar concept.

Although most authors agree that "minor" cycles of this type record small-scale transgressive/regressive pulses, the underlying mechanisms for such trans/regressive episodes are in debate. Wilson (1975) gives an extensive list of possible hypotheses. The generally wide areal distribution of such cycles excludes local autocyclic mechanisms (such as discussed in part I). The alternative allocyclic controls include (1) eustatic fluctuations of absolute sea level, or (2) changes in relative sea level caused by intrabasin tectonics (e.g. differential subsidence).

While glacio-eustatic sea level changes are unlikely in the Triassic, climatically induced fluctuations of global sea level are theoretically conceivable. Cycles in the order of 100,000 years recorded by many authors (see section 3.7.1.) agree well with the 100,000 year climatic cycles predicted by the Milankovitch theory (e.g. Schwarzacher & Fischer, 1982).

Climatically controlled cycles, however, should be correlatable at least within an entire basin, but the Upper Muschelkalk cycles can be safely correlated only within the same paleotectonic unit. Cycles are best developed in the marginal facies belts along the southeastern (on the Moldanubikum, see Fig. 77) and along the northwestern margin (on the Rhenoherzynikum, see logs of Demonconfau, 1982) of the Muschelkalk Basin. Cycles can be well correlated within each of these two structural zones, while cycle correlation between the two margins seems very difficult. In the Upper Muschelkalk of Luxembourg (northwestern basin margin), Demonconfau (1982) recorded 12-15 stacked shallowing-upward cycles, in contrast to more than 30 cycles of the present study area. In the central parts of the Muschelkalk Basin, cycle patterns are different in character and tend to be much more variable than along the more "stable" basin margins, which adds to the difficulty to correlate individual cycles across the entire basin. Possible amalgamation of cycles along basin margins has also to be taken into account.

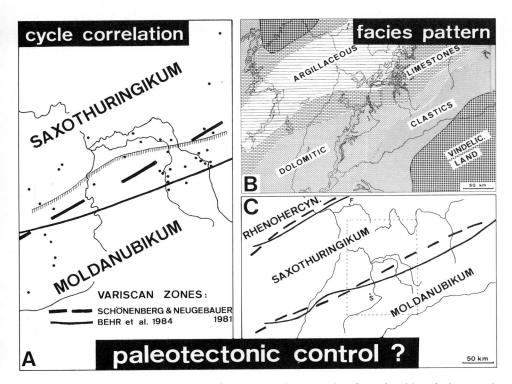

Fig. 77. Evidence for paleotectonic controls in the intracratonic Upper Muschelkalk Basin. A) The line to which coarsening-upward cycles can be readily correlated corresponds fairly well to the boundary between the two Variscan zones of the Moldanubikum and the Saxothuringikum. Dots are measured sections (see Fig. 30). B) and C) The overall facies patterns of the Upper Muschelkalk (B) trace underlying Variscan structural zones (C): marginal facies on Moldanubikum and Rhenoherzynikum respectively, central facies on Saxothuringikum. Note that B) and C) are drawn to the same scale.

The relation of cycle patterns to underlying Variscan structural zones (Fig. 77A) suggests that intrabasin tectonics had an important control on the observed cyclicity. At least two models can be proposed:

Model I assumes a more or less stable global sea level but different subsidence rates of the three structural units. In such a scenario, minor fluctuations in subsidence rates could account for "episodic"deepening on one structural block, causing a minor trans- gression. Since carbonates tend to accumulate at rates greater than average subsidence, they tend to rapidly and repeatedy build up to sea level, resulting in "regressive" shallowing-upward cycles. Because each structural unit behaved differently, the resulting shallowing-upward cycles can be well correlated within each unit. Between units, correlation would be mostly difficult, but some cycles might be present on all structural zones.

Model II assumes similar differential subsidence of the three structural units, but in addition allows for minor changes of global sea level. The interplay between the two mechanisms would produce a different pattern of relative sea level fluctuations, with - as in model I - a different cyclicity on each individual structural block, but a faint correlation between them.

4.2. Hierarchy of cycles

Viewing the Upper Muschelkalk sequence as a whole, two major deposit-ional phases are apparent. Along the southeastern basin margin, there are two distinct periods when shallow-water carbonate "sand" bodies extend far out into the basin, one in the lower and one in the upper part (Fig. 78). These calcareous sand bodies include large amounts of crinoidal debris in the lower part of the Upper Muschelkalk (Hagdorn, 1978), but are dominated by oolitic and shelly carbonates in the upper part of the Upper Muschelkalk. Biostratigraphic correlation by ceratites has shown that crinoidal sand bodies become younger with in-creasing distance from the basin center, both towards the western (Schneider, 1957) and eastern margins of the basin (Hagdorn & Simon, 1981). Similarly, Kleinsorge (1935) reported a shift of the crinoidal limestone facies from the south to the NW in Northern Germany. These patterns clearly mark an overall transgressive shift of facies. The middle part of the Upper Muschelkalk shows the widest aerial extension of marine facies (Kozur, 1974), and thus represents the time of maximum transgression.

In contrast, the upper part of the Upper Muschelkalk has a general re-gressive trend. This is highlighted by the progradation of thick skeletal and oolitic units that extend progressively further into the basin as they become younger. The overall regressive trend is also in-dicated along the southern basin margin, from where a restricted marine to lagoonal dolomitic facies ("Trigonodus-Dolomit") progrades towards the basin center. Similarly, along the northern basin margin, the overall regression is marked by the southward (basinward) pro-gradation of mud- and siltstones that represent prodelta deposits

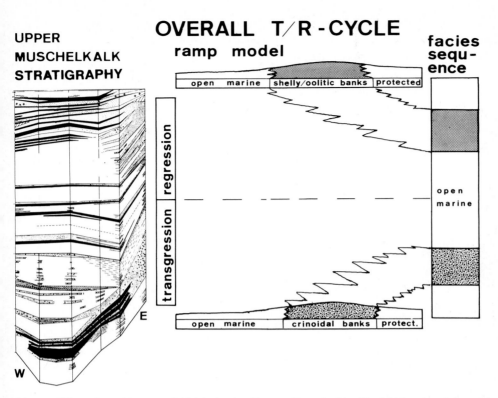

UPPER MUSCHELKALK STRATIGRAPHY

OVERALL T/R-CYCLE
ramp model

facies sequence

open marine | shelly/oolitic banks | protected

regression

transgression

open marine

open marine | crinoidal banks | protect.

W

E

Fig. 78. In the established Upper Muschelkalk lithostratigraphy of SW-Germany (left, after Geyer & Gwinner, 1968), marlstone horizons (black, most prominent in the basin center) can now be regarded as transgressive bases, the massive skeletal/oolitic units (stippled, most prominent at the basin margins) as the regressive tops of minor transgressive/regressive cycles. These asymmetrical minor cycles form the punctuations ("sawtooth pattern") of the overall, nearly symmetrical Upper Muschelkalk transgressive/regressive cycle best recorded along the southeastern basin margin . The ramp depositional model provides a key to understand both the small-scale and the large-scale facies patterns (from Aigner, 1984).

heralding the deltaic conditions of the succeeding Lower Keuper (e.g. Kleinsorge, 1935; Kozur, 1974). Thus, the Upper Muschelkalk sequence as a whole can be regarded as an overall transgressive/regressive cycle.

As schematically indicated in Fig. 78, this overall cycle shows an almost symmetrical development along the SE basin margin and is composed of successive "minor" coarsening-upwards cycles as described above. This means that the overall transgressive and regressive move- ments did not take place gradually, but were punctuated by episodic pulses of a lower order. However, within a sequence of successive minor cycles, the overall trend is often apparent (Fig. 79). During overall regression, for instance, the upper pack- to grainstone parts of stacked minor cycles become more prominent at the expense of the lower mud- to wackestone parts of each cycle, i.e. the cycles become more proximal in appearance.

CYCLE HIERARCHY

Fig. 79. A) The superposition of minor coarsening-upward cycles (blank arrows) becomes apparent in weathered outcrops. Note that the upper, massive (pack- to grainstone) parts of individual cycles are progressively more prominent from bottom to top of this sequence, at the expense of the lower (mud- to wackestone) parts of each cycle. This indicates an overall regressive cycle of a higher order (black arrow). Garnberg, section no. 3 in Fig. 30 and Fig. 74. B) In some quarries, the overall, nearly symmetrical transgressive/regressive cycle of the entire Upper Muschelkalk is obvious. Gundelsheim, section no. 41 in Fig. 30. Scale bars = 5 m.

The approximately 80 m of the whole Upper Muschelkalk cycle encompassed about 5 m.y. using the stratigraphic time scale of Harland et al. (1982) . This is comparable to the order of magnitude duration of "third-order cycles" of Vail et al. (1977). In expanding on the scheme of Vail et al. (1977), "minor", lower-order cycles in the Cretaceous were called "fourth-order cycles" by Ryer (1983), while Busch & Rollins (1984) defined a six-order hierarchy of cycles for the Carboniferous. A similar hierarchy of cycles has been described by Ramsbottom (1979) and by Goodwin & Anderson (1980 a,b) from various Paleozoic epicontinental carbonate sequences.

Since a late Anisian/Ladinian transgression is recognized in the entire Alpine-Mediterranean region (Brandner, 1984) and is also observed along the southern margin of the Tethys onto the Arabian craton (e.g. Druckman et al., 1982), it is possible that a eustatic rise in sea level generated the overall Upper Muschellkalk cycle. Brandner (1984) suggested that these longterm Triassic cycles may be related to increasing spreading rates in the Pacific.

4.3. General context

The Upper Muschelkalk represents but a brief depositional episode within the history of the South-German Basin. This basin represents an "interior sag basin" in the classification of Kingston et al. (1983) and an "intracratonic basin" according to Bally & Snelson´s (1980) scheme. Nevertheless, the Upper Muschelkalk "time slice" sheds light on some general aspects of cratonic basin development and at the same time adresses a variety of open questions.

Cratonic basins are widely taken as tectonically stable units that can be used as sensitive "studios" in which the sea level history is recorded in form of well-defined and widespread sedimentary and faunal cycles (e.g. Cisne et al., 1984). In the Upper Muschelkalk, however, we have observed the influence of small-scale differential paleo-tectonic movements of major structural zones, which have modified, if not dominated the cyclic record of relative sea level fluctuations within different parts of the same basin.

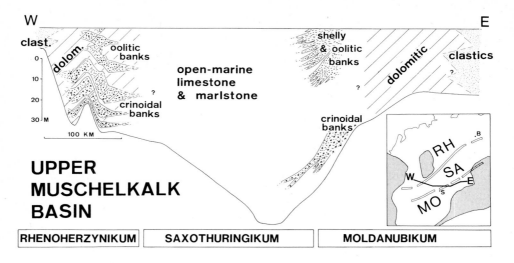

Fig. 80. Strongly generalized W-E section through South-German Upper
Muschelkalk illustrating the influence of Variscan paleotectonic zones
on overall basin organisation. Inserted map shows location of cross-
section in relation to paleogeography (according to Ziegler, 1982),
and to Variscan zones (according to Behr et al., 1984). S = Stuttgart,
B = Berlin. (Based on a compilation of data from Birzer, 1936; Demonconfau, 1982; Geyer & Gwinner, 1968; Haunschild & Ott, 1982, Hary et
al., 1984; Schröder, 1964; Theobald, 1957).

Variscan tectonic elements can be recognized (1) in that the
correlation matrix of minor shallowing-upward cycles is related to
underlying Variscan zones (see Fig. 77), (2) in the NE-strike of local
paleohighs (e.g. "Sierker Schwelle", Demonconfau, 1982) that further
influence local facies patterns, and (3) in that the overall facies
organisation of the Upper Muschelkalk is linked to underlying Variscan
blocks with the marginal facies zones being situated on the Moldanubikum (southeastern margin) and Rhenoherzynikum (northwestern
margin), while the most rapidly subsiding central basinal zone co-
incides with the Saxothuringikum (Figs. 77, 80).

Variscan paleotectonic elements have persisted over a long time (their
influence is still seen in modern topography). They also seem to have
served as "templates" during the history of the South-German Basin. As
already pointed out by Krimmel (1980), the boundary between the
Saxothuringikum and the Moldanubikum has controlled the overall facies
organisation and subsidence patterns of basins from the Zechstein
throughout the Triassic. Generally, subsidence was highest over
Saxothuringian crust, that according to Behr et al. (1984) displays a
multitude of gravimetric and magnetic anomalies as well as electric

and seismic discontinuities, in contrast to the geophysically homogenous Moldanubian crust.

Recent re-evaluations of Variscan zones in terms of plate tectonics
have indicated that "zonal boundaries might be regarded as plate
boundaries (sutures)" (Behr et al., 1984). Subsidence in the South-
German Basin might therefore reflect the reactivation of ancient plate
margin lineaments, similar to what has been observed in other cratonic
basins (cf. Miall, 1984). Sedimentation in the South-German Basin has
started in rapidly subsiding Early Permian troughs and grabens that
follow Variscan strike. According to Ziegler (1982), the subsidence
mechanisms of these basins are still little understood but could in
part involve the cooling of thermal anomalies.

The entire South-German Basin subsided for ca. 160 m.y. with average
subsidence rates of 20 m/m.y.. These values are typical for intracratonic basins that according to Perrodon (1983) characteristically

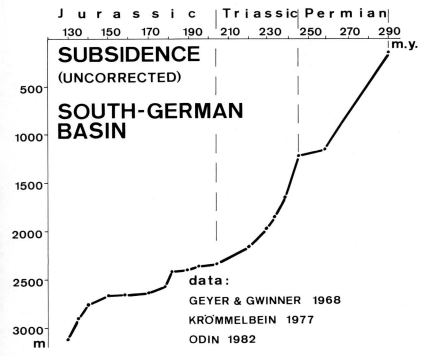

Fig. 81. Uncorrected and very crude subsidence curve for the entire
South-German intracratonic basin. Sediment thicknesses from
Krömmelbein (1977) and Geyer & Gwinner (1968), age determinations from
Odin (1982).

persist for 100-200 m.y. with average subsidence rates of 20 m/m.y. (e.g. Michigan, Illinois, Williston Basin). Fig. 81 shows a very crude subsidence curve for the South-German Basin, uncorrected for compaction, bathymetry, eustacy and isostasy. Nevertheless it shows a characteristic trend of high initial subsidence followed by gradually decreasing rates. This diagram and the limited data base do not allow to categorize the subsidence pattern into either the "thermal contraction model" (Haxby et al., 1976), nor the "crustal stretching model" (McKenzie, 1978). Further work on aspects of "dynamic stratigraphy" and mathematical modelling would be important to fully understand the subsidence mechanisms and the history of the South-German Basin as an example of intracratonic basins.

4.4. Conclusions: basin dynamics (Fig. 82)

The entire South-German Basin can be regarded as the result of a re-activation of plate sutures inherited from the Variscan continental collision. Within this intracratonic basin, the history of the Upper Muschelkalk was largely controlled by an interplay of paleotectonic factors and sea level changes (Fig. 82).

The initial transgression of the Upper Muschelkalk Sea into the German Basin was caused by a (possibly eustatic) rise in sea level during the late Anisian. The almost levelled but gently inclined pre-Upper Muschelkalk topography has resulted in the development of a carbonate ramp depositional system, as it is commoly formed during initial stages of marine transgressions (cf. Wilson, 1975). This pre-existing ramp topography together with the paleolatitude of the basin has allowed storms to play a key role in facies evolution. Small-scale trans/regressive shifts of the carbonate ramp has produced "minor" coarsening-upwards cycles. Vertical stacking of these minor cycles produces the overall transgressive/regressive cycle of the entire Upper Muschelkalk.

Subsidence patterns and overall facies organisation of the Upper Muschelkalk largely trace underlying Variscan structural zones and emphasize paleotectonic controls in this type of intracratonic basin. Different patterns of small-scale trans/regressive cycles on the Molda-nubikum, Saxothuringikum and Rhenoherzynikum indicate different histories of relative sea level fluctuations in each of the Variscan zones. Fluctuations in relative sea level were probably caused mainly by minor differential subsidence of former plates, although super-imposed small-sale eustatic sea level changes cannot be excluded.

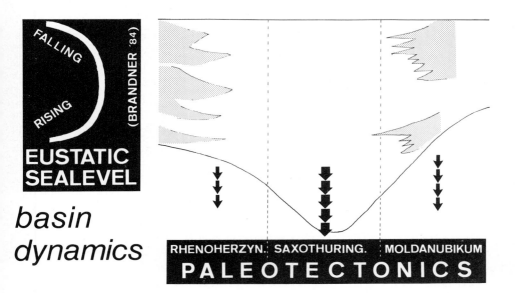

Fig. 82. Strongly schematic summary on the two main factors control-ling intracratonic basin dynamics as exemplified by the Upper Muschel-kalk: (1) the Ladinian rise and fall in (most likely eustatic) sealevel (Brandner, 1984) causing the initial transgression and final regression of the Upper Muschelkalk sea; (2) paleotectonic influences, recorded in different subsidence, facies and cycle patterns over ancient (Variscan) structural zones. Since these probably represent former plates (Behr et al., 1984), epeirogenetic movements in this cratonic basin reflect reactivation of sutures from the Variscan continental collision.

5. DYNAMIC STRATIGRAPHY

CONCLUDING REMARKS

In order to reconstruct aspects of the "dynamic stratigraphy" of an ancient storm depositional system, three levels of stratigraphic sequenceshave been analyzed (Fig. 83):

(1) At the lowest level, different stratification types within the carbonate ramp setting of the Upper Muschelkalk are largely the result of episodic, <u>storm-related processes</u>, whose hydrodynamics can be reconstructed in considerable details. Alongshore moving storms cause distal tempestites with longshore (bipolar) tool marks and gutter cast erosion in deep ramp areas. Due to the Coriolis effect, however, wind stress drove surface water landward causing nearshore water set-up and spillover lobes in shallow ramp skeletal/oolitic banks. The water set-up was compensated by offshore directed bottom return flows, that were responsible for the erosion of storm surge channels through which sediment was funneled offshore to become deposited as proximal tempestites.

(2) At an intermediate level, various types of minor (1-7m) asymmetrical coarsening-upward sequences with distinct trends in lithology, microfacies, paleocurrents and faunas are recorded. These cycles can be explained by repeated small-scale <u>transgressive/ regressive shifts</u> of the carbonate ramp system. Most cycles start with thin but widespread marlstone horizons. Gradual shallowing culminated in a regressive seaward outbuilding of the shallow ramp shoalwater complex, that became subaerially exposed in places. Renewed transgressions are documented by the sudden appearance of a new marlstone horizon.

(3) At a still higher level, vertically stacked minor coarsening-upward cycles form the punctuations and a "sawtooth" pattern within an overall possibly eustatically controlled, transgressive/regressive cycle that comprises the whole of of the Upper Muschelkalk. Many minor cycles provide a basis for litho- and time-stratigraphic correlation over large parts of the basin. More specifically, recognition of this hierarchy and cyclicity allows us to <u>understand</u> the dynamic causes of

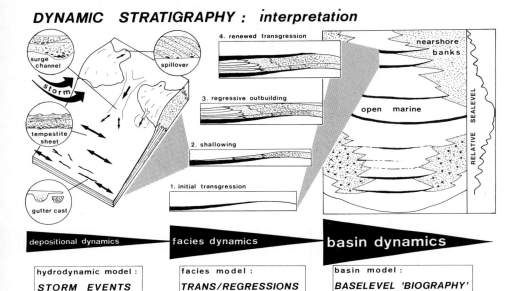

Fig. 83. Summarizing diagram on the reconstruction of dynamic processes based on the analysis of three levels of stratigraphic sequences (cf. Fig 28). (1) <u>Depositional dynamics</u> are dominated by various effects of storms operating in a carbonate ramp setting. (2) <u>Facies dynamics</u> reflect repeated transgressive/regressive shifts of the carbonate ramp generating asymmetrical coarsening-upward cycles. (3) The whole basin shows a hierarchy of cycles: minor, short-term coarsening-upward cycles are superimposed on a major, longer-term trans/regressive cycle. <u>Basin dynamics</u> are controlled by an interplay of eustatic and tectonic factors (from Aigner, 1984).

the purely descriptive lithostratigraphic subdivision so far being used in the Upper Muschelkalk Basin. However, both the overall subsidence and facies patterns and the distribution of minor coarsening-upwards cyles trace underlying Variscan paleotectonic zones. Thus reactivation of former plate sutures from the Variscan continental collision is also a controlling factor for <u>basin dynamics</u>.

In conclusion, the hierarchical stratigraphic analysis outlined here is a simple strategy towards "dynamic stratigraphy", the process-oriented analysis of sedimentary basins. This study attempted to reconstruct some of the processes recorded in an intracratonic storm-dominated basin, but the general principles recognized here should be of value also in other settings.

S O   W H A T ?

At the end such a study, it might be healthy to ask the question "so what ?", which are the general results and what is the broader significance ? This final section is an attempt to highlight very briefly some of the principles recognized in storm depositional systems, their general implications, and possible avenues for future research.

1. Numerous carbonate and terrigenous clastics shallow-marine sequences of continental shelves and epeiric seas are storm-dominated and comprise storm beds (tempestites) interbedded with background sediments. Tempestites from modern and ancient shallow-marine environments show the following "ideal" vertical succession of sedimentary structures: sharp, erosional base, commonly with bi- or multidirectional tool marks, followed by (A) a graded layer of sand or bioclastic material, (B) parallel and/or low-angle lamination (hummocky cross-stratification), (C) wave-ripple lamination and wave ripples, and (D) a mud blanket. Bioturbation commonly starts at tempestite tops and decreases downward in the sequence.

2. The particular association of sedimentary structures suggests combined storm flows with superimposed oscillatory (bi/multidirectional tool marks, wave ripples) and unidirectional components (lateral sediment transport). In spite of superficial similarities, tempestites are significantly different from density-driven event deposits (turbidites).

3. Storm directions can be reconstructed from, and possibly even predicted for, ancient sequences. Storm tracks can be reconstructed from tool marks (prod casts, gutter casts etc.) at tempestite soles and from wave ripple marks on their tops. Predictions are based on present-day latitudinally defined storm systems and global paleogeographical reconstructions, from which paleo-storm systems can be modelled (Klein & Marsaglia, 1983).

4. Onshore storms (e.g. in present-day German Bay) produce onshore oriented wind-drift currents in the surface water, causing landward sediment transport in the nearshore zone (supratidal storm layers, washovers, spillovers). In turn, the coastal water set-up is compen-

sated by offshore directed return flows (gradient currents) in the bottom water causing seaward sediment transport. In such a scenario, on-offshore paleocurrents should prevail.

Longshore storms, in contrast, (e.g. in Atlantic shelf off N-America, or Triassic Muschelkalk) cause alongcoast combined geostrophic bottom flows, driven by the pressure field associated with the set-up (or set-down) of the sea surface against the shore, which is caused by the Coriolis effect (Swift et al., 1983). In the northern hemisphere, the Coriolis effect drives surface water to the right, where landward transport of surface water may again be compensated by coastal down-welling and offshore bottom return flow in the coastal boundary zone. In this case, alongshore paleocurrents are expected in offshore areas, and on-offshore flows in nearshore zones.

5. In a nearshore (proximal) to offshore (distal) direction, the nature of storm stratification varies significantly due to increasing water depth and distance from the coast. Accordingly, individual tempestites show marked proximality trends, expressed by a decrease in bed thickness, grain size, bioclast- and intraclast-content as well as changing sedimentary structures, paleocurrent directions and faunal contents. Fossils can also be used as tracers for storm sediment transport: proximal tempestites commonly contain mixed faunas due to significant lateral sediment influx, while distal tempestites are characterized by parautochthonous assemblages indicating in-situ reworking. On a statistical basis, proximality trends can also be applied to thicker stratigraphic intervals.

6. Proximality trends are a useful tool for paleogeographic reconstructions of storm depositional systems, because they indicate (1) the source areas of storm sands, and (2) paleobathymetric trends. Maps computed from proximality parameters thus help to reconstruct sizes, shapes, geometries and organisations of shallow-marine basins.

7. In basin analysis, vertical tempestite sequences record trans-gressive/regressive fluctuations (e.g. thickening-upward sequences). The packaging of cycles over the whole basin monitors sea level changes and tectonic movements.

8. Due to their geometry (thin sheets, often patchy), individual storm beds cannot be expected to provide significant reservoirs. Amalgamated storm layers, however, forming thicker and laterally more persistant

sand blankets interbedded with shelf muds, are potential hydrocarbon
reservoirs. Such sand bodies are to be expected mainly in regressive
intervals of minor transgressive/regressive cycles, e.g. at the tops
of coarsening-upwards sequences.

9. Storm beds are also useful for a high-resolution stratigraphy.
Individual tempestites provide very sharp time signals, but only on a
local scale, while composite storm-amalgamated or -condensed units,
preferentially at the tops of minor coarsening-upwards cycles, are
often excellent regional markers. They are especially powerful tools
for "event-stratigraphic" correlation, if they are "fingerprinted" by
specific, mostly epibenthic faunas.

10. In paleoecologic analyses, "background" faunal assemblages can be
distinguished from "post-event" faunas. The faunal spectrum of shelly
tempestites, however, might be distorted by (often repeated) storm
reworking and lateral shell influx that has consequences for paleo-
community reconstructions.

11. Storm stratification is also significant in assessing the texture
of the stratigraphical record. The high preservation potential of
event deposits as compared to the long periods of background
conditions leads to a highly episodic or "catastrophic" picture
implying a high degee of stratigraphic incompleteness. This also
limits high-resolution evolutionary studies.

12. Further research should focus on integrated studies along several
avenues, such as :

a) flume experiments and modelling of combined flows under different
hydraulic conditions and with different materials (including bioclastic particles);

b) interdisciplinary cooperation between sedimentologists, marine
geologists, oceanographers, meteorologists and ecologists to monitor
effects and products of present-day storms, including possible
transitions to turbidites (off-shelf transport);

c) further field studies and computer modelling of different scale
cycles composed of storm beds (such as coarsening-up cycles) to better
understand causes for the hierarchy and asymmetric stratigraphic
expression of cycles in storm depositional systems;

d) further examples of integrated basin analysis and "dynamic stratigraphy" in storm depositional systems, combined with mathematical
simulation of controlling factors, such as subsidence, sea level
changes, paleolatitude and paleo-storm regimes. The ultimate goal
would be the establishment of "basin models" that enable us to predict
the evolution of a particular sedimentary basin.

LITERATURE

AGER, D.V.(1974): Storm deposits in the Jurassic of the Moroccan High Atlas. - Palaeogeogr., Palaeoclimatol., Palaeoecol., 15: 83-93.

AGER, D.V. (1981): The nature of the stratigraphical record. - Halsted Press, New York, 122 pp. (second edition)

AGER, D.V. & WALLACE, P. (1970): The distribution and significance of trace fossils in the uppermost Jurassic rocks of the Boulonnais, Northern France. - Geol. J., Spec. Issue No. 3: 1-18.

AHR, W.M. (1973): The carbonate ramp: an alternative to the shelf model. - Trans. Gulf Coast Assoc. Geol. Soc., 23rd Ann. Conv.: 221-225.

AIGNER, T. (1977): Schalenpflaster im Unteren Hauptmuschelkalk bei Crailsheim (Württ., Trias, mol) - Stratinomie, Ökologie, Sedimentologie. - N. Jb. Geol. Paläont., Abh., 153: 193-217.

AIGNER, T. (1979): Schill-Tempestite im Oberen Muschelkalk (Trias, SW-Deutschland). - N. Jb. Geol. Paläont., Abh., 157: 326-343.

AIGNER, T. (1982): Calcareous tempestites: storm-dominated stratification in Upper Muschelkalk limestones (Middle Trias, SW-Germany). In: G. EINSELE & A. SEILACHER (Eds.), Cyclic and event stratification: 180-198. - Springer, Berlin, Heidelberg, New York. (1982a)

AIGNER, T. (1982): Event-stratification in nummulite accumulations and in shell beds from the Eocene of Egypt. In: G. EINSELE & A. SEILACHER (Eds.), Cyclic and event stratification: 248-262. - Springer, Berlin, Heidelberg, New York. (1982b)

AIGNER, T. (1983): Facies and origin of nummulitic buildups: an example from the Giza Pyramids Plateau (Middle Eocene, Egypt). - N. Jb. Geol. Paläont., Abh., 166: 347-368.

AIGNER, T. (1984): Dynamic stratigraphy of epicontinental carbonates, Upper Muschelkalk (M. Triassic), South-German Basin. - N. Jb. Geol. Paläont., Abh., 169: 127-159.

AIGNER, T., HAGDORN, H. & MUNDLOS, R. (1978): Biohermal, biostromal and storm-generated coquinas in the Upper Muschelkalk. - N. Jb. Geol. Paläont., Abh.:, 157: 42-52.

AIGNER, T. & FUTTERER, E. (1978): Kolk-Töpfe und -Rinnen (pot and gutter casts) im Muschelkalk - Anzeiger für Wattenmeer ? - N. Jb. Geol. Paläont., Abh., 156: 285-304.

AIGNER, T. & REINECK, H.-E. (1982): Proximality trends in modern storm sands from the Helgoland Bight (North Sea) and their implications for basin analysis. - Senckenbergiana marit., 14: 183-215.

AIGNER, T. & REINECK, H.-E. (1983): Seasonal variation of wave-base on the shoreface of the barrier island Norderney, North Sea. - Senckenbergiana marit., 15: 87-92.

AITGEN, J.D. (1967): Classification and environmental significance of cryptalgal limestones and dolomites, with illustrations from the Cambrian and Ordovician of southwestern Alberta. - J. sed. Petr., 37: 1163-1178.

ALDINGER, H. (1928): Beiträge zur Stratigraphie und Bildungsgeschichte des Trochitenkalks im nördlichen Württemberg und Baden. - Thesis Univ. Tübingen.

ALLEN, J.R.L. (1982): Sedimentary structures: their character and physical basis. - Developm. Sedimentol. 30 A and B. Elsevier, Amsterdam, A: 593pp, B: 663pp.

ALLEN, J.R.L. (1984): Some general physical implications of storms and their relevance to problems of storm sedimentation. - Brit. Sed. Res. Group Meetg. Storm Sed., Cardiff, Abstr.: 3.

ALLEN, P.A. (1981): Some guidelines in reconstructing ancient sea conditions from wave ripple marks. - Mar. Geol., 43: M59-M67.

ALLEN, P.A. (1984): Reconstruction of ancient sea conditions with an example from the Swiss Molasse. - Mar. Geol., 60: 455-473.

ANDREE, K. (1916): Wesen, Ursachen und Arten der Schichtung. - Geol. Rdsch., 6: 351-397.

ANDERSON, E.J. (1972): Sedimentary structure assemblages in transgressive and regressive calcarenites. - 24th Int. Geol. Congr. Montreal, Sect. 6: 369-378.

AUSICH, W.I. & BOTTJER, D.J. (1982): Tiering in suspension-feeding communites on soft substrata throughout the Phanerozoic. - Science, 216: 173-174.

AUST,H. (1969): Lithologie, Geochemie und Paläontologie des Grenzbereiches Muschelkalk-Keuper in Franken. - Abh. Naturwiss. Ver. Würzburg, 10: 1-155.

BACHMANN, G.H. (1973): Die karbonatischen Bestandteile des Oberen Muschelkalks (Mittlere Trias) in Südwest-Deutschland und ihre Diagenese. - Arb. Inst. Geol. Paläont. Univ. Stuttgart, N.F. 68: 1-99.

BACHMANN, G.H. (1979): Bioherme der Muschel Placunopsis ostracina v. SCHLOTHEIM und ihre Diagenese. - N. Jb. Geol. Paläont., Abh., 158: 381-407.

BACHMANN, G.H. & GWINNER, M.P. (1971): Nordwürttemberg. - Slg. geol. Führer, Bd. 54, 168 pp. Gebr. Borntraeger, Berlin, Stuttgart.

BAGNOLD, R.A. (1966): An approach to the sediment transport problem from general physics. - U.S. Geol. Surv. Prof. Paper 422-I: 37 pp.

BALL, M.M. (1967): Carbonate sand bodies of Florida and the Bahamas. - J. sed. Petr., 37: 556-591.

BALL, M.M., SHINN, E.A. & STOCKMANN, K.W. (1967): The geological effects of Hurricane Donna in South Florida. - J. Geology, 75: 583-597.

BALLY, A.W. & SNELSON,S. (1980): Realms of subsidence. In: A.D. MIALL (Ed.), Facts and principles of world petroleum occurrences. - Can. Soc. Petrol. Geol., Mem., 6: 9-94.

BARRELL, J. (1917): Rhythms and the measurements of geological time. - Bull. Geol. Soc. Am., 28: 745-904.

BARRETT, P.J. (1964): Residual seams and cementation in Oligocene shell calcarenites, Te Kuiti Group. - J. sed. Petr., 34: 524-531.

BARTHEL, K.W. (1974): Black pebbles, fossil and recent, on and near coral islands. - Proc. 2nd Int. Coral Reef Symp., 2. Great Barrier Reef Comm.: 395-399.

BASAN, P.B. (1973): Aspects of sedimentation and development of a carbonate bank in the Barracuda Keys, South Florida. - J. sed. Petr., 43: 42-53.

BAUD, A. (1976): Les terriers de Crustacés decapodes et l'origine de certains faciès du Trias carbonaté. - Eclogae geol. Helv., 69: 415-424.

BEHR, H.J., ENGEL,W., FRANKE, W., GIESE, P. & WEBER, K. (1984): The Variscan Belt in Central Europe: main structures, geodynamic implications, open questions. - Tectonophysics, 109: 15-40.

BIRZER, F. (1936): Eine Tiefbohrung durch das mesozoische Deckgebirge in Fürth in Bayern. - Zbl. Min. etc., 1936,(B): 425-433.

BLENDINGER, W. (1985): Carbonate deposition, strike-slip tectonics and igneous activity in the Ladinian of the central Dolomites, Italy. - Thesis Univ. Tübingen.

BLOOS, G. (1982): Shell beds in the lower Lias of Southern Germany - facies and origin. In: G. EINSELE & A. SEILACHER (Eds.), Cyclic and event stratification: 223-239. - Springer, Berlin, Heidelberg, New York.

BOGACZ,K., DZULYNSKI, S., GRADZINSKI, P. & KOSTECKA, A. (1968): Origin of crumpled limestone in the Middle Triassic of Poland. - Rocs. Pols. Tow. Geol., 38: 385-394.

BOURGEOIS, J. (1980): A transgressive shelf sequence exhibiting hummocky stratification: the Cape Sebastian Sandstone (Upper Cretaceous), Southwestern Oregon. - J. sed. Petr., 50: 681-702.

BRANDNER, R. (1984): Meeresspiegelschwankungen und Tektonik in der Trias der NW-Tethys. - Jb. Geol. B.-Anst., 126: 435-475.

BRENCHLEY, P.J., NEWALL, G. & SANISTREET. J.G. (1979): A storm surge origin for sandstone beds in an epicontinental platform sequence, Ordovician, Norway. - Sediment. Geol., 22: 185-217.

BRENNER, R.L. & DAVIES, D.K. (1973): Storm-generated coquinoid sand-stone: genesis of high-energy marine sediments from the Upper Jurassic of Wyoming and Montana. - Bull. Geol. Soc. Am., 84: 1685-1697.

BRETT, C.E. (1983): Sedimentology, facies and depositional environment of the Rochester Shale (Silurian, Wenlockian) in Western New York and Ontario. - J. sed. Petr., 53: 947-971.

BRINKMANN, R. (1930): Über die Schichtung und ihre Bedingungen. - Fortschr. Geol. Paläont. XI, H. 35: 187-219.

BRÜDERLIN, M. (1969): Beiträge zur Lithostratigraphie und Sediment-petrographie des Oberen Muschelkalks im südwestlichen Baden-Württem-berg. Teil I: Lithostratigraphie. - Jber. Mitt. oberrh. geol. Ver., N.F. 51: 125-158.

BRÜDERLIN, M. (1970): Beiträge zur Lithostratigraphie und Sediment-petrographie des Oberen Muschelkalks im südwestlichen Baden-Württem-berg. Teil II: Sedimentpetrographie. - Jber. Mitt. oberrh. geol. Ver., N.F. 52: 175-209.

BRUNNER, H. & SIMON, T. (1985): Lithologische Gliederung von Profilen aus dem Oberen Muschelkalk im nördlichen Baden-Württemberg anhand der natürlichen Gamma-Strahlungsintensität der Gesteine. - Jber. Mitt. oberrh. geol. Ver., N.F. 67.

BUSCH, R.M. & ROLLINS, H.B. (1984): Correlation of Carboniferous strata using a hierarchy of transgressive-regressive units. - Geology 12: 471-474.

CACCHIONE, D.A. & DRAKE, D.E. (1982): Measurements of storm-generated bottom stresses on the continental shelf. - J. geophys. Res., 87: 1952-1960.

CAIN, J.D.B. (1968): Aspects of the depositional environment and paleo-ecology of crinoidal limestones. - Scott. J. Geol., 4: 191-208.

CANT, D.J. (1984): Development of shoreline-shelf sand bodies in a Cretaceous epeiric sea deposit. - J. sed. Petr., 54: 541-556.

CAROZZI, A.V. & SODERMAN, J.G.W. (1962): Petrography of some Mississippian (Borden) crinoidal limestones at Stobo, Indiana. - J. sed. Petr., 32: 397-414.

CISNE, J.L. GILDNER, R.F. & RABE, B.D. (1984): Epeiric sedimentation and sea-level: synthetic ecostratigraphy. - Lethaia, 17: 267-288.

COOK, D.O. (1970): The occurrence and geological work of rip currents off Southern California. - Mar. Geol. 9: 173-186.

COOK, D.O. & GORSLINE, D.S. (1972): Field observations of sand transport by shoaling waves. - Mar. Geol., 13: 31-55.

CREAGER, J.S. & STERNBERG, R.W. (1972): Some specific problems in understanding bottom sediment distribution and dispersal on the continental shelf. In: D.J.P. SWIFT, D.B. DUANE & O.H. PILKEY (Eds.), Shelf sediment transport: process and pattern: 447-449. Dowden, Hutchinson & Ross, Stroudsburg, Pa.

DEMONFAUCON, A. (1982): Le Muschelkalk supérieur de la Vallee de la Moselle Grand Duché de Luxembourg. - Thesis, Univ. Dijon, 206 pp.

DEUTSCHES HYDROGRAPHISCHES INSTITUT (1975): Nordsee-Handbuch, östlicher Teil. - Hamburg.

DIXON, O.A., NARBONNE, G.M. & JONES, B. (1981): Event correlation in Upper Silurian rocks of Sommerset Island, Canadian Arctic. - Bull. Am. Assoc. Petr. Geol., 65: 303-311.

DÖRJES, J. & HERTWECK, G. (1975): Recent biocoenoses and ichnocoenoses in shallow-marine environments. In: R. FREY (Ed.), The study of trace fossils: 459-491. - Springer, Berlin, Heidelberg, New York.

DOTT, R.H. Jr. (1983): Episodic sedimentation - how normal is average? How rare is rare ? Does it matter ? - J. sed. Petr., 53: 5-23.

DOTT, R.H. Jr. & BOURGEOIS, J. (1982): Hummocky stratification: significance of its variable bedding sequences. - Bull. Geol. Soc. Am., 93: 663-680.

DRUCKMAN, Y., HIRSCH, F. & WEISSBROD, T. (1982): The Triassic of the southern margin of the Tethys in the Levant and its correlation across the Jordan Rift Valley. - Geol. Rdsch., 71: 919-936.

DURINGER, P. (1982): Sédimentologie et paléoécologie du Muschelkalk supérieur et de la Lettenkohle (Trias Germanique) de l'est de la France. Diachronie des faciès et reconstructions des paléoenvironments. - Thesis, Univ. Strasbourg, 96 pp.

DURINGER, P. (1984): Tempetes et tsunamis: des depots de vagues de haute energie intermittente dans le Muschelkalk superieur (Trias germanique) de l'est de la France. - Bull. Soc. geol. France, 1984, (7), t. XXVI, no. 6: 1177-1185.

EBANKS, W.J. & BUBB, J.N. (1975): Holocene carbonate sedimentation, Matecumbe Keys tidal bank, South Florida. - J. sed. Petr., 45: 422-439.

EDER, W. (1982): Diagenetic redistribution of carbonate, a process in forming limestone/marl alternations (Devonian and Carboniferous, Rheinisches Schiefergebirge, W. Germany). In: G. EINSELE & A. SEILACHER (Eds.), Cyclic and event stratification: 98-112. - Springer, Berlin, Heidelberg, New York.

EINSELE, G. & SEILACHER, A. (Eds.) (1982): Cyclic and event stratification. - Springer, Berlin, Heidelberg, New York, 536pp.

ENOS, P. (1983): Shelf. In: P.A. SCHOLLE, D.G. BEBOUT & C.H. MOORE (Eds.), Carbonate depositional environments. - Am. Ass. Petr. Geol., Mem., 33: 267-295.

ENOS, P. & PERKINS, R.D. (1977): Quaternary sedimentation in South Florida. - Geol. Soc. Am., Mem. 147, 198 pp.

FARROW, G.E. (1966): Bathymetric zonation of Jurassic trace fossils from the coast of Yorkshire, England. - Palaeogeogr., Palaeoclimatol., Palaeoecol., 2: 103-151.

FLOOD, R.D. (1981): Distribution, morphology, and origin of sedimentary furrows in cohesive sediment, Southampton Water. - Sedimentology, 28: 511-529.

FORRISTALL, G.Z., HAMILTON, R.C. & CARDONE, V.J. (1977): Continental shelf currents in tropical storm Delia: observations and theory. - J. phys. Oceanogr., 7: 532-546.

FÜRSICH, F.T. (1973): Thalassinoides and the origin of nodular limestone in the Corallian Beds (Upper Jurassic) in Southern England. - N. Jb. Geol. Paläont., Mh., 1973: 136-156.

FÜRSICH, F.T. (1974): Ichnogenus Rhizocorallium. - Paläont. Z., 48: 16-28.

FÜRSICH, F.T. (1975): Trace fossils as environmental indicators in the Corallian of England and Normandy. - Lethaia, 8: 151-172.

FRANK, M. (1937): Paläogeographischer Atlas von Südwestdeutschland. - Mitt. geol. Abt. württ. statist. Landesamt, 17: 1-111.

FUTTERER, E. (1978): Studien über die Einregelung, Anlagerung und Einbettung biogener Hartteile im Strömungskanal. - N. Jb. Geol. Paläont., Abh., 156: 87-131.

GADOW, S. & REINECK, H.-E. (1969): Ablandiger Sandtransport bei Sturmfluten. - Senckenbergiana marit., 1: 63-78.

GEBELEIN, C.D. (1977): Dynamics of recent carbonate sedimentation and ecology, Cape Sable, Florida. - E.J. Bill, Leiden, 120 pp.

GEYER, O.F. & GWINNER, M.P. (1968): Einführung in die Geologie von Baden-Württemberg. - E. Schweizerbart´sche Verlagsbuchhandlung, Stuttgart, 228 pp.

GIENAPP, H. (1972): Wellenmessungen im Seegebiet der Piep (Deutsche Bucht). - Helgol. Wiss. Meeresunters., 23: 261-267.

GIENAPP, H. (1973): Strömungen während der Sturmflut vom 2. November 1965 in der Deutschen Bucht und ihre Bedeutung für den Sedimenttransport. - Senckenbergiana marit., 5: 135-151.

GIENAPP, H. & TOMCZAK, G. (1968): Strömungsmessungen in der Deutschen Bucht bei Sturmfluten. - Helgol. Wiss. Meeresunters., 17: 94-107.

GINSBURG, R.N., BERNARD, H.A., MOODY, R.A. & DAIGLE, E.E. (1966): The Shell method of impregnating cores of unconsolidated sediment. - J. sed. Petr., 36: 1118-1125.

GINSBURG, R.N. & JAMES, N.P. (1974): Spectrum of Holocene reef-building communities in the Western Atlantic. - Sedimenta IV: 7.2-7.22.

GOLDRING, R. & BRIDGES, P. (1973): Sublittoral sheet sandstones. - J. sed. Petr., 43: 736-747.

GOLDRING, R. & AIGNER, T. (1982): Scour and fill: the significance of event separation. In: G. EINSELE & A. SEILACHER (Eds.), Cyclic and event stratification: 354-362. - Springer, Berlin, Heidelberg, New York.

GOODWIN, P.W. & ANDERSON, E.J. (1980): Punctuated aggradational cycles: a general hypothesis of stratigraphic accumulation. - Geol. Soc. Am., Abstr. with Progr., 12: 436 (1980 a).

GOODWIN, P.W. & ANDERON, E.J. (1980): Application of the PAC hypothesis to limestones of the Helderberg Group. - Soc. Econ. Pal. Min., Eastern Section, Field Conf. Guidebook 1980: 1-32. (1980b)

GRAY, D.J. & BENTON, M.J. (1982) Multidirectional paleocurrents as indicators of shelf storm beds. In: G. EINSELE & A. SEILACHER (Eds.), Cyclic and event stratification: 350-353. - Springer, Berlin, Heidelberg, New York.

GWINNER, M.P. (1970): Revision der lithostratigraphischen Nomenklatur im Oberen Hauptmuschelkalk des nördlichen Baden-Württemberg. - N. Jb. Geol. Paläont., Mh., 1970: 77-87.

GWINNER, M.P. & HINKELBEIN, K. (1976): Stuttgart und Umgebung. - Slg. geol. Führer, 61, 148 pp. Gebr. Borntraeger, Berlin, Stuttgart.

HAGAN, G.M. & LOGAN, B.W. (1974): Development of carbonate banks and hypersaline basins, Shark Bay, Western Australia. In: B.W. LOGAN et al. (Eds.), Evolution and diagenesis of Quaternary carbonate sequences, Shark Bay, Western Australia. - Am. Assoc. Petr. Geol., Mem. 22: 61-139.

HAGDORN, H. (1978): Muschel/Krinoiden-Bioherme im Oberen Muschelkalk (mol, Anis) von Crailsheim und Schwäbisch Hall (Südwestdeutschland). - N. Jb. Geol. Paläont., Abh., 156: 31-86.

HAGDORN, H. (1982): The "Bank der kleinen Terebrateln" (Upper Muschelkalk, Triassic) near Schwäbisch Hall (SW-Germany) - a tempestite condensation horizon. In: G. EINSELE & A. SEILACHER (Eds.), Cyclic and event stratification: 263-285. - Springer, Berlin,

Heidelberg, New York.

HAGDORN, H. (1985): Immigration of crinoids into the German Muschel-kalk Basin. In: U. BAYER & A. SEILACHER (Eds.), Sedimentary and Evolutionary cycles: 237-254. - Springer, Berlin, Heidelberg, New York.

HAGDORN, H. & MUNDLOS, R. (1982): Autochthonschille im Oberen Muschelkalk (Mitteltrias) Südwestdeutschlands. - N. Jb. Geol. Paläont., Abh., 162: 332-351.

HAGDORN, H. & MUNDLOS, R. (1983): Aspekte der Taphonomie von Muschelkalk-Cephalopoden. Teil 1: Siphozerfall und Füllmechanismus. - N. Jb. Geol. Paläont., Abh., 166: 369-403.

HAGDORN, H. & SIMON, T. (1981): Oberer Muschelkalk. In: H. BRUNNER et al., Schichtenfolge und geologische Bedeutung der Thermalwasserbohrung Aalen 1. - Jh. Ges. Naturkde. Württ., 136: 56-61.

HÄNTZSCHEL, W. (1936): Die Schichtungsformen rezenter Flachmeer-Ablagerungen. - Senckenbergiana leth., 18: 316-356.

HAMBLIN, A.P. & WALKER, R.G. (1979): Storm-dominated shallow-marine deposits: the Fernie-Kootenay (Jurassic) transition, southern Rocky Mountains. - Can. J. Earth Sci., 16: 1673-1690.

HARBOUGH, J.W. (1957): Mississippian bioherms of Northeast Oklahoma. - Bull. Am. Assoc. Petr. Geol., 41: 2530-2544.

HARDIE, L. A. (Ed.) (1977): Sedimentation on the modern carbonate tidal flats of Northwest Andros Island, Bahamas. - Johns Hopkins Univ. Studies in Geology, no. 22, 202 pp.

HARDIE, L.A. & GINSBURG, R.N. (1977): Layering: the origin and environmental significance of lamination and thin bedding. In: L.A. HARDIE (Ed.), Sedimentation on the modern carbonate tidal flats of Northwest Andros Island, Bahamas. - Johns Hopkins Univ. Studies in Geology, no. 22: 50-123.

HARLAND, W.B. et al. (1982): A geologic time scale. - Cambridge Earth Science Ser., Univ. Press Cambridge, 131pp.

HARMS, J.C., SOUTHARD, J.B., SPEARING, D.R. & WALKER, R.G. (1975): Depositional environments interpreted from primary sedimentary structures and stratification sequences. - Soc. Econ. Pal. Min., Short Course no. 2, 161 pp.

HARY, A., BOCK, H., DITTRICH, D. & WAGNER, J.F. (1984): Trias in Becken- und Randfazies im Luxemburger Gutland. - Jber. Mitt. oberrh. Geol. Ver., N.F. 66: 85-94.

HAUNSCHILD, H. & OTT, W.-O. (1982): Profilbeschreibung, Stratigraphie und Paläogeographie der Forschungsbohrung Dinkelsbühl 1001. - Geologica Bavarica, 83: 5-55.

HAXBY, W.F., TURCOTTE, D.L. & BIRD, J.M. (1976): Thermal and mechanical evolution of the Michigan Basin. - Tectonophysics, 36: 57-75.

HAYES, M.O. (1967): Hurricanes as geological agents, south Texas coast. - Bull. Am. Assoc. Petr. Geol., 51: 937-942.

HELLER, P.L., KOMAR, P.D. & PEVEAR, D.R. (1980): Transport processes in ooid genesis. - J. sed. Petr., 50.: 943-952

HINE, A.C. (1977): Lily Bank, Bahamas: history of an active oolite sand shoal. - J. Sed. Petr., 42: 1554-1581

HOLLOWAY, S. (1983): The shell-detrital calcirudites of the Forest Marbel Formation (Bathonian) of southwest England. - Proc. Geol. Assoc., 94: 259-266.

HOWARD, J.D. & REINECK, H.-E. (1981): Depositional facies of high-energy beach-to-offshore sequence: comparison with low-energy sequence. - Bull. Am. Assoc. Petrol. Geol., 65: 807-830.

HUNTER, R.E. & CLIFTON, H.E. (1982): Cyclic deposits and hummocky cross-stratification of probable storm origin in Upper Cretaceous rocks of the Cape Sebastian area, Southwestern Oregon. - J. sed. Petr., 52: 127-143.

IRWIN, M.L. (1965): General theory of epeiric clear water sedimentation. - Bull. Am. Assoc. Petr. Geol., 49: 445-459.

JAMES, N.P. (1980): Shallowing-upward sequences in carbonates. - Geoscience Canada, Repr. Ser. 1: 109-119.

JENKYNS, H.C. (1971): Speculations on the genesis of crinoidal limestones in the Tethyan Jurassic. - Geol. Rdsch., 60: 471-488.

JENKYNS, H.C. (1974): Origin of red nodular limestones (Ammonitico Rosso, Knollenkalk) in the Mediterranean Jurassic: a diagenetic model. - Spec. Publ. int. Ass. Sediment., 1: 249-271.

JOHNSON, R.G. (1957): Experiments on the burial of shell. - J. Geology, 65: 527-535.

JOHNSON, H.D. (1978): Shallow siliciclastic seas. In: H.G. READING (Ed.), Sedimentary environments and facies: 207-258. - Blackwell, Oxford, Edinburgh, Boston, Melbourne.

JONES, B., OLDERSHAW,A.E. & NARBONNE, G.M. (1979): Nature and origin of rubbly limestone in the Upper Silurian Read Bay Formation of Arctic Canada. - Sedim. Geol., 24: 227-252.

KÄLIN, O. (1980): Schizosphaerella punctulata DEFLANDRE & DANGEARD: wall ultrastructure and preservation in deeper-water carbonate sediments of the Tethyan Jurassic. - Eclogae geol. Helv., 73: 983-1008.

KAZMIERCZAK, J. & PSZCZOLKOWSKI, A. (1969): Burrows of enteropneusta in Muschelkalk (Middle Triassic) of the Holy Cross Mountains, Poland. - Acta Palaeont.Pol., 14: 299-315.

KAZMIERCZAK, J. & GOLDRING, R. (1978): Subtidal flat-pebble conglomerate from the Upper Devonian of Poland: a multi-provenant high-energy product. - Geol. Mag., 115: 359-366.

KEARY, R. & KEEGAN, B.F. (1975): Stratification by infauna debris: a structure, a mechanism and a comment. - J. sed. Petr., 45: 128-131.

KELLEY, V.C. (1956): Thickness of strata. - J. sed. Petr., 26: 289-300.

KENNEDY, W.J. & GARRISON, R.E. (1975): Morphology and genesis of hardgrounds and nodular chalks in the Upper Cretaceous of Southern England. - Sedimentology, 22: 311-386.

KIDWELL, S.M. & JABLONSKI, D. (1983): Taphonomic feedback. Ecological consequences of shell accumulation. In: M.J.S. TEVESZ & P.L. McCALL (Eds.), Biotic interactions in recent and fossil benthic communities: 195-248. - Plenum Publ. Corp.

KINGSTON, D.R., DISHROON, C.P. & WILLIAMS, P.A. (1983): Global basin classification system. - Bull. Am. Assoc. Petr. Geol., 67: 2175-2193.

KLEIN, G. de V. (1971): A sedimentary model for determining paleotidal range. - Bull. Geol. Soc. Am., 82: 2585-2592.

KLEINSORGE, H. (1935): Paläogeographische Untersuchungen über den Oberen Muschelkalk in Nord- und Mitteldeutschland.. - Mitt. Geol. Staatsinst. Hamburg, 15: 57-106.

KOLP, O. (1958): Sedimentsortierung und Umlagerung am Meeresboden durch Wellenwirkung. - Petermanns Geogr. Mitt., 102: 173-178.

KOMAR, P.D., NEUDECK, R.H. & KULM, L.D. (1972): Observations and significance of deep water oscillatory ripple marks on the Oregon continental shelf. In: D.J.P. SWIFT, D.B. DUANE & O.H. PILKEY (Eds.), Shelf sediment transport: processes and pattern: 601-619. - Dowden, Hutchinson & Ross, Stroudsburg, Penn.

KOZUR, H. (1974): Biostratigraphie der germanischen Mitteltrias. - Freib. Forschungsh., C 280: 7-56.

KREISA, R.D. (1981): Storm-generated sedimentary structures in subtidal marine facies with examples from the Middle and Upper Ordovician of Southwest Virginia. - J. sed. Petr., 51: 823-848.

KRIMMEL, V. (1980): Epirogene Paläotektonik zur Zeit des Keupers (Trias) in Südwest-Deutschland. - Arb. Inst. Geol. Paläont. Univ. Stuttgart, N.F. 76: 1-74.

KRÖMMELBEIN, K. (1977): Brinkmanns Abriß der Geologie. Historische Geologie. - Enke-Verlag Stuttgart, 400 pp.

KUMAR, N. & SANDERS, J.E. (1976): Characteristics of shoreface storm deposits: modern and ancient examples. - J. sed. Petr., 46: 145-162.

LAPORTE, L.F. (1969): Recognition of a transgressive carbonate sequence within an epeiric sea: Helderberg Group (Lower Devonian) of New York State. - Soc. Econ. Pal. Min., Spec. Publ., 14: 98-119.

LINCK, O. (1965): Stratigraphische, stratinomische und ökologische Betrachtungen zu Encrinus liliiformis LAMARCK. - Jh. geol. L.-A. Baden-Württ., 7: 123-148.

LOGAN, B.W., REZAK, R. & GINSBURG, R.N. (1964): Classification and environmental significance of algal stromatolites. - J. Geology, 72: 68-83.

LOGAN, B.W. et al. (1969): Carbonate sediments and reefs, Yukatan shelf, Mexico. - Am. Assoc. Petr. Geol., Mem, 11: 1-196.

LOGAN,B.W. & SEMENIUK, V. (1976): Dynamic metamorphism, processes and products in Devonian carbonate rocks, Canning Basin, Western Australia. - Geol. Soc. Australia, Spec. Publ. No. 6, 138pp.

LOMBARD, A. (1978): Sedimentology. In: R.W. FAIRBRIDGE & J. BOURGEOIS, (Eds.) Encyclopedia of Sedimentology: 703-707. - Dowden, Hutchinson & Ross, Stroudsburg, Pa.

LOREAU, J.P. & PURSER, B.H. (1973): Distribution and ultrastructure of

Holocene ooids in the Persian Gulf. In: B.H. PURSER (Ed.), The Persian Gulf: 279-328. - Springer, Berlin, Heidelberg, New York.

LOVELL, J.P.B. (1970): The palaeogeographical significance of lateral variation in the ratio of sandstone to shale and other features in the Aberystwyth Grids. - Geol. Mag., 107: 147-158.

MARKELLO, J.R. & READ, J.F. (1981): Carbonate ramp-to-deeper shale transitions of an Upper Cambrian intrashelf basin, Nolichucky Formation, Southwest Virginia, Appalachians. - Sedimentology, 28: 573-597.

MARKELLO, J.R. & READ, J.F. (1982): Upper Cambrian intrashelf basin, Nolichucky Formation, Southwest Virginia Appalachians. - Bull. Am. Assoc. Petr. Geol., 66: 860-878.

MARSAGLIA, K.M. & KLEIN, G. de V. (1983): The paleogeography of Paleozoic and Mesozoic storm depositional systems. - J. Geology, 91: 117-142.

MATTHEWS, R.K. (1984): Dynamic stratigraphy. - Prentice-Hall, New Yersey, 2nd ed., 489 pp. (1st ed. 1974).

MAYER, G. (1952): Lebensspuren von Bohrorganismen aus dem Unteren Hauptmuschelkalk (Trochitenkalk) des Kraichgaues. - N. Jb. Geol. Paläont., Mh., 1952: 440-456.

MAYER, G. (1955): Eine interessante Schichtfläche aus dem Mittleren Hauptmuschelkalk von Bruchsal. - Beitr. naturk. Forsch. SW-Deutschl., 14: 114-118.

McCAVE, I.N. (1971): Sand waves in the North Sea off the coast of Holland. - Mar. Geol., 10: 199-225.

McKEE, E.D. & WEIR, G.W. (1953): Terminology for stratification and cross-stratification. - Bull. Geol. Soc. Am., 64: 381-390.

McKENZIE, D.P. (1978): Some remarks on the development of sedimentary basins. - Earth planet. Sci. Lett., 40: 25-32.

MEHL, J. (1982): Die Tempestit-Fazies im Oberen Muschelkalk Südbadens. - Jh. geol. L.-A. Baden-Württ., 24: 91-109.

MERKI, P. (1961): Der Obere Muschelkalk im östlichen Schweizer Jura. - Eclogae geol. Helv., 54: 137-219.

MIALL, A. (1984): Principles of sedimentary basin analysis. - Springer, Berlin, Heidelberg, New York, 490 pp.

MILLER, M.C. & KOMAR, P.D. (1977): The development of sediment threshold curves for unusual environments (Mars) and for inadequately studied materials (foram sands). - Sedimentology, 24: 709-721.

MORTON, R.A. (1981): Formation of storm deposits by wind-forced currents in the Gulf of Mexico and the North Sea. - Spec. Publ. int. Ass. Sediment., 5: 385-396.

MULLINS, H.T. et al. (1980): Nodular carbonate sediment on Bahamian slopes: possible precursors to nodular limestones. - J. sed. Petr., 50: 117-131.

NELSON, C.H. (1982): Modern shallow-water graded sand layers from storm surges, Bering Shelf: a mimic of Bouma-sequences and turbidite systems. - J. sed. Petr., 52: 537-545.

NELSON, C.H. & NIO, S.D. (Eds.)(1982): The northeastern Bering Shelf: new perspectives of epicontinental shelf processes and depositional products. - Spec. Issue Geol. en Mijnbouw, 61: 1-114.

NIO, S.D., SHÜTTENHELM, R.T.E. & WEERING, Tj.C.E. van (Eds.) (1981): Holocene marine sedimentation in the North Sea Basin. - Spec. Publ. int. Ass. Sediment., 5, 515 pp.

NUMMEDAL, D. et al. (1980): Geologic response to hurricane impact on low-profile Gulf Coast barriers. - Trans. Gulf Coast Assoc. Geol. Soc., XXX: 183-195.

ODIN, S. (1982): Numerical dating in stratigraphy. - John Wiley & Sons, 1040 pp.

PAUL, W: (1936): Der Hauptmuschelkelk am südöstlichen Schwarzwald. - Mitt. bad. geol. Landesamt, 11: 123-146.

PEMBERTON, S.G. & FREY, R.W. (1982): Trace fossil nomenclature and the Planolites-Palaeophycus dilemma. - J. Paleont., 56: 843-881.

PERKINS, R.D. & ENOS, P. (1968): Hurricane Betsy in the Florida-Bahama area - geologic effects and comparison with Hurricane Donna. - J. Geology, 76: 710-717.

PERRODON, A. (1983): Geodynamique des bassins sédimentaires et systèmes pétroliers. - Bull. Centre Rech. Expl. Prod. Elf-Aquitaine, 7: 645-676.

PERYT, T.M. (1980): Structure of "Sphaerocodium kokeni Wagner" a Girvanella oncoid from the Upper Muschelkalk (Middle Triassic) of Württemberg, SW-Germany. - N. Jb. Geol. Paläont., Mh., 1980: 293-302.

PURSER, B.H. (Ed.) (1973): The Persian Gulf. - Springer, Berlin, Heidelberg, New York, 471 pp.

PURSER, B.H. & SEIBOLD, E. (1973): The principal environmental factors influencing Holocene sedimentation and diagenesis in the Persian Gulf. In: B.H. PURSER (Ed.), The Persian Gulf: 1-9. - Springer, Berlin, Heidelberg, New York.

RAMSBOTTOM, W.H.C. (1979): Rate of transgression and regression in the Carboniferous of NW Europe. - J. geol. Soc. London, 136: 147-153.

READ, J.F. (1982): Carbonate platforms of passive (extensional) continental margins: types, characeristics and evolution. - Tectonophysics, 81: 195-212 (1982a).

READ, J.F. (1982): Carbonate platform models: tectonic controls, facies, depositional processes and diagenesis. - Am. Assoc. Petrol. Geol. Fall Educ. Conf., Denver: 1-36 (1982b).

REES, E.I.S., NICHOLAIDOU, A. & LASKARIDOU, P. (1977): The effects of storms on the dynamics of shallow water benthic associations. In: B.F. KEEGAN, P. O'CEDIGH & P.J.S. BOADEN (Eds.), Biology of benthic organisms. - Proc. 11th Europ. Symp. Mar. Biol., 465-474, Pergamon Press, Oxford.

REIF, W.-E. (1971): Zur Genese des Muschelkalk-Keuper-Grenzbonebeds in Südwestdeutschland. - N. Jb. Geol. Paläont., Abh., 139: 369-404.

REIF, W.-E. (1982): Muschelkalk/Keuper bone-beds (Middle Triassic, SW-Germany) - storm condensation in a regressive cycle. In: G. EINSELE

& A. SEILACHER (Eds.), Cyclic and event stratification: 299-325. - Springer, Berlin, Heidelberg, New York.

REINECK, H.-E. (1962): Die Orkanflut vom 16. Februar 1962. - Nat. u. Mus., 92: 151-172.

REINECK, H.-E. (1963): Sedimentgefüge im Bereich der südlichen Nordsee. - Abh. Senckenb. naturf. Ges., 5o5: 1-138.

REINECK, H.-E. (1969): Zwei Sparkerprofile südöstlich Helgoland. - Nat. u. Mus., 99: 9-14.

REINECK, H.-E. (1977): Natural indicators of energy level in recent sediments: the application of ichnology to a coastal engineering problem. - Geol. J., Spec. Issue 9: 265-272.

REINECK, H.-E. & SINGH, I.B. (1972): Genesis of laminated sand and graded rhythmites in storm layers of shelf mud. - Sedimentology, 18: 123-128.

REINECK, H.-E. & SINGH, I.B. (1980): Depositional sedimentary environments. - Springer, Berlin, Heidelberg, New York, 549pp (2nd Ed.).

REINECK, H.-E., GUTMANN, W.F. & HERTWECK, G. (1967): Das Schlickgebiet südlich Helgoland als Beispiel rezenter Schelfablagerungen. - Senckenbergiana leth., 48: 219-275.

REINECK, H.-E., DÖRJES, J., GADOW, S. & HERTWECK, G. (1968): Sedimentologie, Faunenzonierung und Faziesabfolge vor der Ostküste der inneren Deutschen Bucht. - Senckenbergiana leth., 49: 261-3o9.

REINHARDT, J. & HARDIE, L.A. (1976): Selected examples of carbonate sedimentation, Lower Paleozoic of Maryland.- Maryland Geol. Surv., Guidebook no. 5: 1-53.

RICHTER, R. (1929): Gründung und Aufgaben der Forschungsstelle für Meeresgeologie "Senckenberg" in Wilhelmshaven. - Nat. u. Mus., 59: 1-3o.

RICKEN, W. (1985): Diagenetische Bankung - Sedimentologie von Kalk/Mergel-Wechselfolgen. - Thesis Univ. Tübingen.

RUHRMANN, G. (1971): Riff-ferne Sedimentation unterdevonischer Krinoidenkalke im Kantabrischen Gebirge (Spanien). - N. Jb. Geol. Paläont., Mh., 1971: 231-248.

RUPPEL, S.C. & WALKER, K.R. (1982): Sedimentology and distinction of carbonate buildups: Middle Ordovician, East Tennessee. - J. sed. Petr., 52: 1055-1071.

RYER, T.A. (1983): Transgressive-regressive cycles and the occurrence of coal in some Upper Cretaceous strata of Utah. - Geology, 11: 207-210.

SADLER, P.M. (1982): Bed-thickness and grain size of turbidites. - Sedimentology, 29: 37-51.

SCHÄFER, K.A. (1973): Zur Fazies und Paläogeographie der Spiriferina-Bank (Hauptmuschelkalk) im nördlichen Baden-Württemberg. - N. Jb. Geol. Paläont., Abh., 143: 56-110.

SCHÄFER, W. (1970): Aktuopaläontologische Beobachtungen. 9. Faunenwechsel. - Senckenbergiana marit., 2: 85-102.

SCHNEIDER, E. (1957): Beiträge zur Kenntnis des Trochitenkalkes des Saarlandes und der angrenzenden Gebiete. - Ann. Univ. Sarav., 6: 185-257.

SCHÖNENBERG, R. & NEUGEBAUER, J. (1981): Einführung in die Geologie Europas. - Rombach, Freiburg, 340 pp. (4th ed.).

SCHRÖDER, B. (1967): Fossilführende Mittlere Trias im Ries. - Geol. Bl. NO-Bayern, 17: 44-47.

SCHWARZACHER, W. & FISCHER, A.G. (1982): Limestone-shale bedding and perturbations of the Earth's orbit. In: G. EINSELE & A. SEILACHER (Eds.), Cyclic and event stratification: 72-95. - Springer, Berlin, Heidelberg, New York.

SEILACHER, A. (1960): Strömungsanzeichen im Hunsrückschiefer. - Notizbl. hess. L.-Amt Bodenforsch., 88: 88-106.

SEILACHER, A. (1967): Bathymetry of trace fossils. - Mar. Geol., 5: 413-428.

SEILACHER, A. (1982): Distinctive features of sandy tempestites. In: G. EINSELE & A. SEILACHER (Eds.), Cyclic and event stratification: 333-449. - Springer, Berlin, Heidelberg, New York.

SHACKLEY, S.E. & COLLINS, M. (1984): Variations in sublittoral sediments and their associated macro-infauna in response to inner shelf processes; Swansea Bay, U.K. - Sedimentology, 31: 793-804.

SHEA, J.H. (1982): Twelve fallacies of uniformitarianism. - Geology, 10: 455-460.

SHINN, E.A. (1983): Tidal flat. In: P.A. SCHOLLE, D.B. BEBOUT & C.H. MOORE (Eds.), Carbonate depositional environments. - Am. Assoc. Petr. Geol., Mem. 33: 171-210.

SHROCK, R.R. (1948): Sequence in layered rocks. - Mc Graw-Hill, New York, 507 pp.

SKUPIN, K. (1969): Lithostratigraphische Profile aus dem Trochitenkalk des Neckar-Jagst-Kocher-Gebietes. - Jber. Mitt. oberrh. geol. Ver., N.F. 51: 87-118.

SKUPIN, K. (1970): Feinstratigraphische und mikrofazielle Untersuchungen im Unteren Hauptmuschelkalk des Necker-Jagst-Kocher-Gebietes. - Arb. geol.-paläont. Inst. Univ. Stuttgart, N.F. 63: 1-173.

SMITH, J.D. & HOPKINS, T.S. (1972): Sediment transport on the continental shelf of Washington and Oregon in light of recent current measurements. In: D.J.P. SWIFT, D.B. DUANE & O.H. PILKEY (Eds.), Shelf sediment transport: Process and Pattern: 143-180. - Dowden, Hutchinson & Ross, Stroudsburg, Pa.

SMITH, A., HURLEY, A.M. & BRIDEN, J. (1981): Phanerozoic paleocontinental world maps. - Cambridge Univ. Press, Cambridge, 102 pp.

SOMERVILLE, J.D. (1979): A cyclicity in the early Brigatian (D2) limestones east of Clwydian Range, North Wales, and its use in correlation. - Geol J., 14: 69-86.

STERNBACH, L.R. & FRIEDMANN, G.M. (1984): Deposition of ooid shoals marginal to the Late Cambrian Proto-Atlantic (Japetus) Ocean in New York and Alabama; influence on the interior shelf. - Soc. Econ. Pal. Min., Core Workshop no. 5: 2-19.

STERNBERG, R.W. & LARSEN, L.H. (1972): Frequency of sediment movement on the Washington continental shelf: a note. - Mar. Geol., 21: M37-M47.

STRASSER, A. & DAVAUD, E. (1983): Black pebbles of the Purbeckian (Swiss and French Jura): lithology, geochemistry and origin. - Eclogae geol. Helv., 76: 551-580.

SUNDBORG, A. (1967): Some aspects of fluvial sediments and fluvial morphology. I. General views and graphic methods. - Geogr. Ann.: 333-343.

SUNDQUIST, B. (1982): Paleobathymetric interpretation of wave ripple-marks in a Ludlovian grainstone sequence of Gotland. - Geol. Fören. Stockh. Förenhandl., 102: 157-166.

SWIFT, D.J.P., FIGUEIREDO, A.G. Jr., FREELAND, G.L. & OERTEL, G.F. (1983): Hummocky cross-stratification and megaripples: a geological double standard ? - J. sed. Petr., 53: 1295-1317.

THEOBALD, N. (1952): Stratigraphie du Trias moyen dans le sud-ouest de L'Allemagne et le nord-est de la France. - Ann. Univ. Sarav., Saar-brücken, 64 pp.

TRUSHEIM, F. (1931): Versuche über Transport und Ablagerung von Mollusken. - Senckenbergiana, 13: 124-139.

TURMEL, R.J. & SWANSON, R.G. (1976): The development of Rodriguez Bank, a Holocene mudbank in the Florida Reef Tract. - J. sed. Petr., 46: 497-518.

VAIL, P.R. MITCHUM, R.M. & THOMPSON, S. III. (1977): Seismic strati-graphy and global changes of sea level, Part 4: Global cycles of relative changes of sea level. - Am. Ass. Petr. Geol., Mem., 26: 83-97.

VOIGT, E. (1975): Tunnelbau rezenter und fossiler Phoronoidea. - Paläont. Z., 49: 135-167.

VOLLRATH, A. (1938): Zur Stratigraphie und Bildung des Oberen Haupt-muschelkalks in Mittel- und Westwürttemberg. - Mitt. Min.-Geol. Inst. TH Stuttgart, 33: 69-80.

VOLLRATH, A. (1955): Zur Stratigraphie des Hauptmuschelkalks in Württemberg. - Jh. geol. L.-A. Baden-Württ., 1: 79-168. (1955a)

VOLLRATH, A. (1955): Zur Stratigraphie des Trochitenkalks in Baden-Württemberg. - Jh. geol. L.-A. Baden-Württ., 1: 169-189. (1955b).

VOLLRATH, A. (1955): Stratigraphie des Oberen Hauptmuschelkalks (Schichten zwischen Cycloides-Bank und Spiriferina-Bank) in Baden-Württemberg. - Jh. geol. L.-A. Baden-Württ., 1: 190-216. (1955c)

VOLLRATH, A. (1957): Zur Entwicklung des Trochitenkalks zwischen Rhein-tal und Hohenloher Ebene. - Jh. geol. L.-A. Baden-Württ., 2: 119-134.

VOLLRATH, A. (1958): Beiträge zur Paläogeographie des Trochitenkalks in Baden-Württemberg. - Jh. geol. L.-A. Baden-Württ., 3: 181-194.

VOLLRATH, A. (1970): Ein vollständiges Profil des Oberen Muschelkalks und ein neues Mineralwasser bei Ummenhofen, Gemeinde Untersontheim, Landkreis Schwäbisch Hall. - Jber. Mitt. oberrh. geol. Ver., N.F. 52: 133-148.

WAGNER, G. (1913): Beiträge zur Kenntnis des oberen Hauptmuschelkalks in Elsass-Lothringen. - Cbl. Min. Geol. Paläont., 1913: 551-589 (1913a).

WAGNER, G. (1913): Beiträge zur Stratigraphie und Bildungsgeschichte des Oberen Hauptmuschelkalks und der Unteren Lettenkohle in Franken. - Geol. Paläont. Abh., N.F. 12: 275-452 (1913b).

WALKER, R. (1967): Turbidite sedimentary structures and their relationship to proximal and distal depositional environments. - J. sed. Petr., 37: 25-43.

WALKER, R. (1970): Review of the geometry and facies organisation of turbidites and turbidite-bearing basins. In: L. LAJOIE (Ed.), Flysch sedimentology in North America. - Spec. Pap. geol. Assoc. Can., 7: 219-251.

WALKER, R. (180): Facies and facies models. - Geoscience Canada, Reprint Ser., 1: 1-7. (1980a)

WALKER, R. (1980): Shallow-marine sands. - Geoscience Canada, Reprint Ser., 1: 75-89. (1980b)

WANLESS, H.R. (1969): Sediments of Biscayne Bay - distribution and depositional history. - Inst. Mar. Atmos. Sci. Univ. of Miami, Tech. Rept. 62-6, 260pp.

WANLESS, H.R. (1970): Influence of pre-existing bedrock topography on bars of "lime" mud and sand, Biscayne Bay, Florida. - Bull. Am. Ass. Petr. Geol., Abstr., 54: 875.

WANLESS, H.R. (1976): Intracoastal sedimentation. In: D.J. STANLEY & D.J.P. SWIFT (Eds.), Marine sediment transport and environmental management: 221-239. - John Wiley & Sons.

WANLESS, H.R. (1978): Storm-generated stratigraphy of carbonate mud banks, South Florida. - Geol. Soc. Amer., Abstr. with Progr., 10: 512.

WANLESS, H.R. (1979): Role of physical sedimentation in carbonate-bank growth. - Bull. Am. Assoc. Petr. Geol., Abstr., 63: 547. (1979a)

WANLESS, H.R. (1979): Limestone response to stress: pressure solution and dolomitization. - J. sed. Petr., 49: 437-462. (1979b)

WANLESS, H.R. (1981): Fining-upwards sedimentary sequences generated in seagrass beds. - J. sed. Petr., 51: 445-454.

WANLESS, H.R., BURTON, E.A. & DRAVIS, J. (1981): Hydrodynamics of carbonate fecal pellets. - J. sed. Petr., 51: 27-36.

WARZESKI, R.E. (1976): Storm sedimentation in the Biscayne Bay region. - Biscayne Bay Symp., Univ. of Miami Sea Grant Spec. Rpt. No. 5: 34-38.

WENGER, R. (1957): Die germanischen Ceratiten. - Palaeontographica, A, 108: 57-129.

WETZEL, A. (1979): Ökologische und stratigraphische Bedeutung biogener Gefüge in quartären Sedimenten am NW-afrikanischen Kontinentalrand. - Meteor-Forsch.-Ergebn., Reihe C, No. 34: 1-47.

WHITACKER, J.H. McD. (1973): "Gutter casts", a new name for scour- and fill-structures: with examples from the Llandoverian of Ringerike and

Malmöya, southern Norway. - Norsk geol. Tidesskrift, 53: 4o3-417.

WILSON, J.L. (1975): Carbonate facies in geologic history. - Springer, Berlin, Heidelberg, New York, 471 pp.

WILSON, J.L. & JORDAN, C. (1983): Middle shelf. In: P.A. SCHOLLE, D.G. BEBOUT & C.H. MOORE (Eds.), Carbonate depositional environments. - Am.Ass. Petr. Geol., Mem, 33: 297-343.

WIRTH, W. (1957): Beiträge zur Stratigraphie und Paläogeographie des Trochitenkalkes im nordwestlichen Baden-Württemberg. - Jh. geol. L.-A. Baden-Württ., 2: 135-173.

WRIGHT, M.E. & WALKER, R.G. (1981): Cardium Formation (Upper Cretaceous) at Seebe, Alberta - storm-transported sandstones and conglomerates in shallow-marine depositional environments below fair-weather wave-base. - Can. J. Earth Sci., 18: 795-809.

WUNDERLICH, F. (1979): Die Insel Mellum (südliche Nordsee). Dynamische Prozesse und Sedimentgefüge. I. Südwatt, Übergangszone und Hochfläche. - Senckenbergiana marit., 11: 59-113.

WUNDERLICH, F. (1983): Sturmbedingter Sandversatz vor den Ostfriesischen Inseln und im Gebiet des Großen Knechtsandes, Deutsche Bucht, Nordsee. - Senckenbergiana marit., 15: 199-217.

ZIEGLER, P.A. (1982): Triassic rifts and facies patterns in Western and Central Europe. - Geol. Rdsch., 71: 747-772.